'That Important and Necessary Article'

ade
ide
850

l Society
n No. 22

Dundee
1984

ACKNOWLEDGEMENTS

Few people engaged in research in Scottish history can pursue the subject of their enquiry for long without having to call upon the services and assistance of the staffs of the National Library of Scotland and the Scottish Record Office. To them I wish to express my gratitude for help with this and other projects in which I have been engaged in recent years. I am deeply indebted too to Captain Michael Wemyss, of Wemyss Castle, Fife, for allowing me to have access to some of the family's invaluable collection of material relating to both coal mining and salt manufacturing at Dysart, Methil and Wemyss itself. For help and generous hospitality when I was studying this material at Wemyss I should like to express my thanks to Mrs Alice Cunnock. I have benefited from discussions, especially about the technical aspects of the industry and its upstanding remains, with Graham Douglas and John Hume, of the University of Strathclyde, Ian Donnachie of the Open University in Scotland, Colin Martin of the University of St Andrews, Bruce Walker of Duncan of Jordanstone College of Art, Dundee, Geoffrey Stell of the Royal Commission on the Ancient and Historical Monuments of Scotland and Alex Muir of Dundee College of Technology. The Carnegie Trust for the Universities of Scotland have generously funded my travelling costs. The Map and Figure 1 were drawn by Mr J. Ford of the Department of Geography, University of Dundee. My greatest debt however is to Dr Annette Smith, Honorary Editor of the Abertay Historical Society, who has been an endless source of encouragement, without which this publication is unlikely to have appeared.

Finally I should assure my wife Lilian, and children Eilidh and Neil, that I shall in future abandon work during the Christmas and New Year festive period. For their tolerance this time I owe them my sincere thanks.

C.A. Whatley
Department of Modern History
University of Dundee

January 1984

CONTENTS

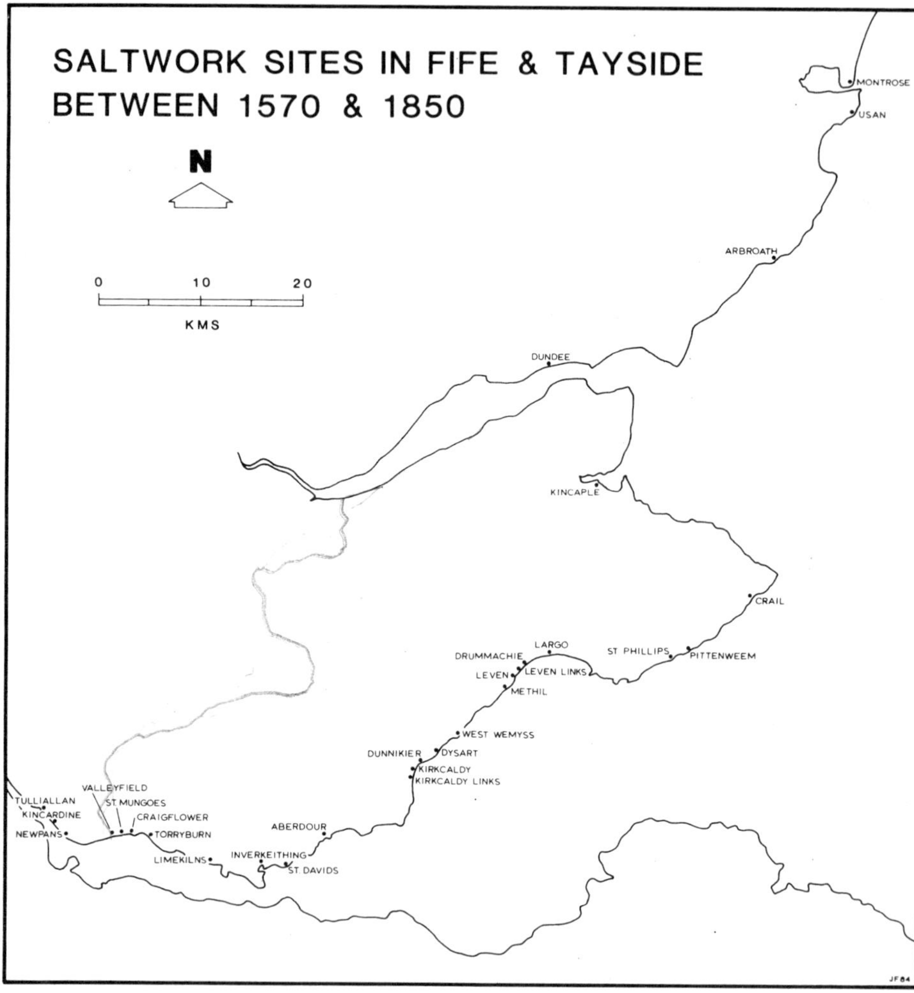
SALTWORK SITES IN FIFE & TAYSIDE
BETWEEN 1570 & 1850
N
0
10
20
KMS
MONTROSE
USAN
ARBROATH
DUNDEE
KINCAPLE
CRAIL
LARGO
ST PHILLIPS
PITTENWEEM
DRUMMACHIE
LEVEN
LEVEN LINKS
METHIL
WEST WEMYSS
DUNNIKIER
DYSART
KIRKCALDY
KIRKCALDY LINKS
ABERDOUR
INVERKEITHING
ST. DAVIDS
LIMEKILNS
TORRYBURN
CRAIGFLOWER
ST. MUNGOES
VALLEYFIELD
TULLIALLAN
KINCARDINE
NEWPANS
JF84

INTRODUCTION

Salt, wrote a contributor to the *Scottish Monthly Magazine* in 1836, was, after corn, 'perhaps the most important of the necessaries of life'. [1] Not only did it add some spice to the oatmeal-based diet which was the lot of most Scots prior to the nineteenth century, but it was also a vital preservative of fish and meat, thereby making human survival through the long winter months both possible and more tolerable. [2] Salt also had some industrial uses, although one seventeenth century critic of the Scottish 'marine' variety claimed that its use in leather preparation made the hide 'fall out in holes'. [3] It was however used as a glaze by potters and as a flux by glassmakers; thus, whilst the existence of salt making capacity at Dundee is little known, that this was operated by the short-lived Dundee Glasswork Company in the 1790s should occasion little surprise.

Fife's* connection with the salt manufacturing industry and its trade is well-known. Contemporaries often remarked upon the deep imprint which it made on the region's appearance, economy and urban life. Gazing over the River Forth towards Fife in the 1630s for example, Sir William Brereton reported that the saltpans were 'innumerable' and 'infinite' in their extent. [4] For some localities salt making was virtually the sole industry; in 1722, Alexander Rait noted that in Tulliallan parish was 'the toun of Kincardin made up mostly of salt pans and those that are employed about them'. [5] He was not only referring to the salters and their assistants but to a wide variety of tradesmen, who obtained at least some proportion of their incomes from tasks performed at the pans. Smiths were involved in building and repairing pans, joiners constructed pan stands, mended doors, and made and repaired carts and other equipment, masons built and rebuilt panhouses, hearths, girnels† and the salters' houses, while tailors could be found making and mending the considerable quantities of 'pocks' [0] in which the salt was carried around the works. In addition carriers were required to bring iron, nails, wood, stone, lime, heather or straw (for thatching) and other materials for the works. There were too a handful of casual and, sometimes, permanent employees who carried out a variety of tasks, such as Betty Adamson at Dysart whose half year's house rent of 12s.6d. was paid in November 1817 'for washing Salt Bags'. [6] At Culross saltwork

* The modern region of Fife. Thus Culross, which during the period under review was in Perthshire, is included in this study.

† granaries, storehouses.

0 sacks

employment may have done much to sustain the local economy; in 1656 Thomas Tucker had noted that there, as in the case of Kincardine, 'salt onely goes out'.[7] Similar circumstances prevailed at Methil and Wemyss, while the establishment of a colliery and saltwork at St Monance was said by David Loch, in 1778, to have 'saved the town' of Pittenweem, causing it to 'flourish more than any on the east coast of Fife'.[8] Later, in 1817, Robert Bald urged that Dysart colliery and its saltwork, whose joint survival was critically dependent upon salt sales, should be carried on with 'vigour', as it was 'the only establishment connected with the Town which produces a circulation of money and the means of Subsistence to a great number of People'.[9]

Rightly, few historians have referred to Fife's coastal collieries without remarking, like Bruce Lenman, that 'nearly all of them had coal-fired salt pans near the shore from which they exported salt'.[10] However it should not be assumed that salt making was a mere offshoot of coal mining - some collieries were opened largely to provide the fuel for salt manufacturing, while many only survived through the seventeenth and much of the eighteenth century owing to the strength of their salt sales. These could be considerable; around half of the salt made and sold in Scotland during the first half of the eighteenth century came from Fife's saltworks.[11]

At first sight the inclusion of Tayside in a study such as this may appear somewhat incongruous. The region after all has never in the past been associated with salt manufacturing and lacked the industry's basic locational requirement, plentiful supplies of cheap coal for firing the pans. Yet, as the map on page 6 shows, saltworks were opened in Tayside, at Montrose and Usan (1794), Dundee (1795) and Arbroath (1814), and formed the southern links of a chain of similar establishments which stretched as far north as Brora in Sutherland.[12] Why this altogether new development in the Scottish salt industry occurred requires explanation. As will be seen, it was a matter of serious concern to Fife's saltmasters, who had previously supplied the greatest proportion of the domestic salt requirements of the inhabitants of Tayside and the coastal areas to the north. Indeed, it was partly to strike a blow at this 'improper monopoly' of the Forth's salt traders that David Scott determined to build a saltwork at Usan.[13] Ultimately Tayside's salt needs were only partially satisfied by the new works, whose combined sales between 1795 and 1798 were just over nine per cent of the total achieved by their Fife counterparts. Nevertheless the local impact of a saltwork could be marked in Tayside too. In 1796, John Leighton, David Scott of Dunninald's factor, stressed the importance of the recently constructed saltpans at Usan, as 'by increasing the population and promoting the Circulation of a little Money among the people of the Village' the value of the estate was raised.[14]

Yet, in spite of the undoubted importance of salt as both an article of consumption and a manufactured commodity, neither the industry nor its trade in Scotland has been the subject of serious investigation. Only one scholarly article devoted to Scottish marine salt making has appeared until recently.[15] It is then largely to fill this major historiographical gap for one part of Scotland at least, that this essay has been written. However, partly because it has been judged important to deal with the subject over a long period, from around 1570 until 1850, and partly as there is so little previous work to draw from (none in the case of Tayside) this paper should not be considered as anything like an attempt to make a definitive statement. It deals systemically with what is known, and incorporates much which is new, but questions about its interpretation remain, and some matters have had to be ignored altogether, owing to limitations of space and time. Thus the scale of investment in the industry, the origins and performance of its entrepreneurs, and saltwork profitability continue to require investigation. Similarly there remain vast untapped archival reserves for historians of individual saltworks and their communities.

What follows is divided into three sections. The first, 'Making Salt: Problems and Processes', is concerned to describe how salt was made and what variations occurred, either between works or over time, in techniques and in the scale of operation. The second, 'Output, Sales and Markets', attempts to describe the course which these three variables followed over the period under review, and also tries to account for the most significant changes which occurred. The final section, 'Salter Serfs', examines the conditions of employment of an occupational group whose experience has been assumed to be identical to the coal miners of Scotland, with whom the salters shared certain legal limitations. Unfortunately, the important question of income levels has had to be ignored. The materials for such a study exist, but practical and methodological problems which the treatment of such a long period raises require that it be dealt with separately. A full, though by no means exhaustive, set of 'Notes', listing the main sources, precedes two appendices which list the area's saltworks in both the second and final decades of the eighteenth century respectively, and give the reader an indication of their relative importance, as judged by salt sales.

CHAPTER 1

Making Salt: Problems and Processes

The manufacture of salt, by boiling sea water in iron pans, remained a rather primitive technological process throughout the period under review. First, sea water was collected in natural rock-cut basins or man-made ponds, widely known as 'bucket pots' although the use of hand buckets to remove water into the pans, the process which gave them their name, became increasingly rare. These were used both as a form of settling tank, wherein mud and other extraneous matter could fall to the bottom and thus be kept out of the pan, and as a reservoir to provide water supplies for the salt workers when the tide was out (see Plate 1). Some slight evaporation also took place. From the bucket pots, the water was transferred, either by hand or pump, through what were originally known as 'pan spouts', that is wooden, or later iron, channels or pipes, into the iron pan itself, situated inside the panhouse. This was usually a stone-built structure with a steeply-pitched turf or heather covered roof.[16] There the water was heated but, prior to boiling, quantities of egg whites or blood were thrown in to form a 'black frothy scum' in which further impurities were gathered before being drawn off by the salters. This done, evaporation proper could begin, with 'small' coal or 'panwood'* being used as the fuel. As many as six tons of coal were required to make one ton of 'marine' salt. It was a lengthy process, lasting at least twenty four hours. During this period the pan was refilled three, four or even five times (depending on the salinity of the sea water) until a 'full' or whole pan of salt had been made. The hot salt was then drawn to the sides of the pan, allowed to drain for a short time and then removed in baskets or 'pocks' or what were also called 'drabs' to the girnel or storehouse. The length of the operation can be seen in the case of James Thomson, a customs official at Usan saltworks. When he applied for eight days leave in August 1798, he felt it necessary to propose that 'his Duty ... be performed in the day-time by Mr Daniel Adamson ... and in the night-time by James Petrie', to ensure that throughout the time when the pans were boiling the salters were discouraged from removing salt illegally.[17]

A brief description such as this however, although essential at this stage, cannot hope to incorporate the wide variety of local circumstances and techniques used. Of necessity it overlooks most of the modest, but nevertheless perceptible, changes in production methods

* panwood or pancoal was the name often given to the small pieces of coal used to fire the pans.

1. Usan, Tayside. Remains of saltwork. This panhouse was converted into an icehouse in the early nineteenth century. Still visible are the chimney, on the seaward wall, a water intake and the bucket pot, pond or reservoir, in the foreground.

which occurred between the sixteenth and the nineteenth centuries. Above all perhaps, it conveys the impression that salt manufacturing was a rather straightforward matter, and by so doing conceals the many problems faced by the salt makers, the identification and elucidation of which may assist us to achieve a fuller understanding of the work process in this little-researched industry.

In 1641, as part of an attempt to induce more regular and sober working practices in both coal mines and saltworks, the Scottish Parliament passed an Act which insisted that colliers and salters worked on six days of the week. For each day they lay idle a fine of 20s. (Scots)* 'or other punishment of ther bodies' could be imposed.[18] While many, if not all, of Scotland's saltmasters may have wished to establish such a regime and increase output from their works, the fact is that frequent interruptions, both regular and irregular, were inherent in the salt-making process. Typical is the example of the pans belonging to the family of Wemyss of Bogie, in Kirkcaldy. There, in the twelve months from October 1720, in only 25 of the 52 possible weeks were all five pans going at once. In less than half (45 per cent) of the total 260 'pan weeks' (i.e. 5 pans x 52) was salt made for a full week.[19] While there was some rationalisation of working methods as well as uneven advances in saltpan technology during the eighteenth and early nineteenth centuries, the pattern of interrupted production continued, if less severely. At Dysart, in the six months from May 1817, in only eleven of the possible 156 'pan weeks' (i.e. 6 pans x 26) was no salt made. However in 63 'pan weeks', or 40 per cent of those worked, production was seriously interrupted for one reason or another.[20] As will be seen later in this pamphlet, some breaks, at these and other works, can be attributed to workers exercising their collective or individual 'leisure preference'[21] and volunteering to cease work for a time. Nevertheless, the majority of stoppages appear to have been caused by factors over which there was little or no human control, although in some areas improvements in production could be, and were, effected over time.

The salter's calendar was punctuated by two regular periods when salt could not be made. At what were usually monthly intervals, each pan was stopped and cooled down while 'paddling' or 'paidling' took place, an operation which lasted one or two days. Almost half of the breaks in production at Dysart during the period mentioned above were attributable to this. The problem was caused by the sides of the pan becoming 'crusted over with some sort of matter' and forming a hard scale,

* worth $^1/_{12}$ of the value of its English equivalent.

which, if not removed, caused the iron pan plates to burn. The salters knocked off what was sometimes called 'stone scratch' (but which was calcium sulphate) with iron picks. Failure to carry out paddling caused holes to appear in the pan plates, which then had to be replaced at the owner's expense. Presumably it was to avoid this that the 2nd Earl of Wemyss, in the mid-seventeenth century, paid his salters 'paidling' money of 24s. (Scots) monthly, in addition to providing them with two and a half 'dozen'† of coals, two-fifths of which were to be used for re-starting the pan while 'the other is allowed him [the salter] and his man for the work, and all the salt they can make off that dozen and a half'.[22] A longer but far less frequent stoppage occurred at intervals of roughly two years, when the 'beiting' took place. The pans and panstands, heavily scaled and burnt during long periods of constant use, had to be stripped down, repaired and replaced where necessary, and then rebuilt. The time taken to complete this work varied and was dependent upon factors such as the extent of the damage, the availability of materials, notably iron for plates and nails, and whether or not suitable smiths and other tradesmen could be found. Thus, at Dysart in 1751, while there was no shortage of smiths, their payments varied from ten and a half days work at both the Craig and East pans to twenty days at the West pan.[23] Even more extreme was the case of James Foord, a salter at Bogie works, whose pan was 'at beit' for the best part of seven weeks during 1720, for reasons which the salt grieve unfortunately failed to record.[24] For smiths living in the vicinity of a saltwork, pan repairing could form a significant proportion of their work. For example, at Dysart between February and September 1715, a total of ten smiths were employed, five of whom spent more than ten and a half days, equivalent perhaps to two working weeks, at the pans for at least five of those months.[25] Some other saltmasters preferred to contract with a single smith who would be responsible for all of the pan work.

For the salters a lengthy 'beiting' period could lead to considerable hardship, as their money incomes were based upon the quantities of salt they produced. At some work, Wemyss and Methil in the 1650s for example, they were paid an allowance of either a firlot of meal or £3 per week, 'this being my Lord's pleasure ... to give the meal or the money' noted Earl David in his diary. Others were less inclined to support their

† a quantity of coal. Accurate conversion of pre-nineteenth century Scottish weights and measures into standard English measures is still a difficult task, in spite of at least one recent attempt to standardise some of the more commonly used measures; see R.E. Zupko, 'The weights and measures of Scotland before the Union', *Scottish Historical Review*, LVI (1977), 119-145.

temporarily unproductive workers in this way and, although the evidence is restricted to a scatter of rather brief account books and papers, it does appear that the Wemyss practice, not universally adopted in the seventeenth century, was rarely followed in the eighteenth. Instead, payments to salters during periods of enforced idleness were made for specific labouring jobs, such as assisting the smiths at the pans, or 'mending the Walls of the East Bucket pot', for which Sir Robert Henderson of Fordel's salters were paid 6s. in April 1755.[26]

Less easy to cope with were the many unpredictable factors which halted salt production. By far the most serious of these were interruptions in the supply of coal to the pans. Indeed the exhaustion of coal supplies in a given locality could lead to the abandonment of the pans, as occurred at Torry and Tulliallan.[27] More often however fuel shortages were temporary. When Thomas Tucker surveyed the industry in 1656, it was 'failure of coale' which was the main impediment to weekly production.[28] At the same time, at Methil and Wemyss, meal or cash allowances, payable when there was a 'want of coals', were an integral part of the salter's contract.[29] While the problem may have become less serious as more pits were sunk and coal output increased during the eighteenth century, thereby producing more waste or 'pancoal', it was still in evidence in the early nineteenth century. In just under one-third of the weeks in which salt production was interrupted or stopped altogether at Dysart in the six months from May to November 1817, a coal shortage was to blame. As the colliery was then in a poor condition however this may well be a rather extreme case.

Partly because of the nature of the saltmaking process and partly owing to their exposed situation, the panhouses themselves required frequent repairs which could only be effected when the fires were out. Indeed it was the continuous heat required to make salt which caused the furnaces, hearths and 'lums' (or chimneys) to crack and crumble. Rebuilding could take up to a fortnight. At Kirkcaldy's Bogie pans in the early eighteenth century the salters were required to carry out this work themselves,[30] although this was not common. At most works masons were hired and then assisted by the salter and his servant whose stonework was being repaired. A serious storm could halt production for some time. The best-known was that of 1625 which was said to have destroyed most of the saltpans (and some collieries) on the Forth, including those at Culross.[31] Another, in January 1752, caused severe structural damage at Dysart; quarriers had to be hired to bring stones to 'make up the breatches thatt the Storme Meade upon the Sallt pans', while a small army of sixteen masons, assisted by the salters and others, was employed in rebuilding the panhouses.[32] Less serious, as work might only be stopped for a matter of hours or at most a day, was the frequency with which the wind blew seaweed and other unwanted material into the bucket pots. Salt grieves had to make numerous small payments to their salters, or other persons, for removing the waste. Thus at Bogie in January 1727 expenses of 4s. (Scots) were incurred for 'takeing the sea-warr out of the mid-pan pott'[33] whilst much later, in 1817, it seems to have been common for a party of four salters to be paid 1s. each for cleaning out one of their number's bucket pots at Dysart.

Rain, especially where this was accompanied by strong east winds, was also able to halt salt production as the fresh water from swollen streams and rivers, unable to disperse itself, reduced the saline content of the sea water in the vicinity of the pans, thereby rendering salt production impossible or prohibitively expensive. The problem of the 'suddaine coming downe of the freshes' had been noted by Thomas Tucker,[34] and its presence continued to be felt throughout the period. In the week beginning 24 May 1817 for example, William Murray, a salter at Dysart's 'Middle' pan, was able to make only three instead of the usual four 'fulls' of salt as he was 'Stopt by E. Wind', while George Skinner at the nearby 'Craiggie's' pan found himself short by a similar quantity owing to bad weather late in October.[35]

Of the disrupting influences which can be readily categorised, illness amongst the salters and their servants is probably the least significant, although the fact that it is identifiable may, sometime in the future, allow comparisons to be made with other occupational groups. At present it is not possible to judge how far sickness amongst salt workers was occupationally derived, although the payment of a quarterly fee to 'Dr Law for Medicine & attendance to Salters' at Dysart in 1817 and 1818 points to the certain existence of some work-related medical problems.[36] The extent to which this was true, however, is a matter for speculation; all we can do here is briefly record how often illness interfered with production. At Bogie, in the twelve months from October 1720, three salters were unable to work for a total of six weeks, with Andrew Hound, who lost three weeks production, being the worst affected. Here, and in the seventeenth century at Methil and Wemyss, the sick were not left to fend for themselves entirely and some assistance was given by their master. In February 1723, Patrick Davidson, a salter at Bogie, was paid £1.16s. (Scots) 'Dureing the time of his Sickness' while some time later £3 (Scots) was disbursed from saltwork funds for his 'Coffine'. From the limited amount of evidence currently available there is no suggestion that the problem diminished over time. At Dysart in

1817, for instance, just under seven per cent of stoppages came under headings such as 'Unwell' or 'servant badly' in the works' oncost book.

Finally, in addition to the more common reasons why salt production was so often interrupted, there were a number which are best classified in a 'miscellaneous' category. In October 1720 for instance, James Foord, at Bogie saltworks, lost one day's output because of the 'downfall' of his pan. More serious was the case of his fellow salter, John Hound, who lost a complete week's income from salt making in November owing to 'his servent flitting'. After 1713, when a tax became payable on sales of Scottish salt (with certain exemptions), it was not uncommon to find salters jailed for 'running' the commodity which they made. As early as July 1714, the grieve at David St Clair's Dysart saltwork had to pay £2.8s. (Scots) for the costs of the '8 Days the Two Salters was in Kinghorn prison'[37] while in 1724, at the nearby Bogie pans, 6s. (Scots) was paid for 'relieving Jo. Ffoord from prisone put in by Custom house officers'. In March 1753, production at the extensive Wemyss saltworks was almost entirely halted when five salters there were incarcerated by Kirkcaldy's magistrates for allegedly 'clandestinely selling ... Salt to Salt Cadgers'.[38]

Few if any of the features of the salt making process so far described would not have been recognised by anyone who was acquainted with the industry at any time during the period under review. There were however some important differences between the various saltworks, notably in the methods and scale of salt making. Changes occurred in each of these areas during these three hundred years, but none uniformly. Uneven development is perhaps the most striking feature here, even within the restricted confines of two Scottish regions.[39] In order to account for this a number of inter-related factors have to be considered, including inertia, the force of local custom and beliefs about best working practices, the suitability of raw materials, not least the salinity of the sea water and, above all, the degree to which individual saltmasters were concerned with efficiency and the quality of their product.

As has been seen, the salt making process began with the collection and storage of a sufficient quantity of sea water. Filling the 'bucket pot' was a relatively simple matter with water 'coming up with the flood tide through channels or watercourses', a number of which can still be seen, at Usan and St Monance for instance (see plates 2 and 3). Whilst this was a fairly common practice, the numbers and sizes of the bucket pots varied considerably. Traditionally each salt pan had its own bucket pot, from which water was transferred, by the salter, his assistant, or members of his family, by hand bucket, into the pan. However this was

2. Usan, Tayside. Probably Scotland's largest saltwork conduit or water course. From the late 1790s, it also served as a shelter for small fishing vessels.

3. St Monance, Fife. View through the water channel which took water into a large reservoir, which is still visible on site, cut into the rock. The panhouses ranged along the shore above the high-water mark. The windmill was used to pump water from the reservoir into the nine pans.

a slow, laborious and time-consuming process which was evidently being improved upon in the early seventeenth century, perhaps in response to the increased demand for Scottish salt overseas. Thus in 1658, during the early stages of the construction of Methil saltworks, it was considered best to 'make pumps to put the water ... in the pans and not buckets as at Wemyss'. Soon afterwards the hand bucket system at Wemyss must have been abandoned as, on his visit there in 1720, Henry Kalmeter noted that it was a 'windmill that draws up the water'.[40] A further modification was the introduction of a single large water tank in place of the individual bucket pot. This not only made possible the use of a single pump, as in the case of the Wemyss windmill which served eight pans, but also went some way towards eliminating the irritation and expense involved in the upkeep and repair of several smaller 'pots'. A larger tank had the added advantage of allowing the better settlement of the 'heterogeneous ingredients' whose presence so offended contemporaries. From 1772 a single tank or pond, from which water was pumped by a cliff-top windmill into nine pans, was used at St Monance saltworks (also called St Philips and Newark), the 'neatest and best contrived ... upon the coast' according to one observer.[41] The base of the windmill still survives, as can be seen in Plate 3. Similarly at Culross in the late eighteenth century, five pans were served by one 'bucket pot' whose ten and a half foot thick, twelve feet high walls enclosed a collecting area measuring 75 by 60 yards.[42] At Usan, too, a single rock-cut basin, into which the sea flowed through a deep channel, supplied the needs of three saltpans. By far the most impressive arrangement was at St Davids where, some time prior to 1836, a steam engine had been built to pump water into a reservoir large enough to hold 'as much water as is necessary for a week's consumption'.[43] Yet there were still works which utilised the time-honoured methods, as at Dysart less than two decades earlier, when each of the seven pans had its own bucket pot. At the same time Dysart remained Fife's third largest producer of salt.[44]

Equally marked were variations in the critical area of panhouse technique. As in the case of the previous stage of the saltmaking process older methods continued to be used alongside the new. At Bogie and Methil, in the early 1720s, the underpan heating arrangements, whereby a 'strong fire' was situated less than two feet below the underside of the pan, were considered to be crude in comparison with those at Cockenzie saltworks on the south bank of the Forth. There the fire was five feet below the pans and the heat more evenly distributed, thereby reducing the frequency and expense of 'beiting'; the Bogie pans it was alleged, 'when going to beet are all miserable Scorch'd or ... burnt'.[45] While the salters there may have been unwilling to alter established working practices, partly in the belief that they produced a more potent salt, at the nearby Leven saltworks remarkably sophisticated salt manufacturing methods had been introduced by 1720. Not only were 'brander' or 'bander' pan used there, which were fired by simple underpan furnaces, as opposed to the method used at the more common 'sole' pans where coal was simply thrown onto the ground underneath the pan itself, but also the water entering the large iron pans was pre-heated in lead pans suspended on top of them.[46] Whilst the use of such advanced fuel-efficient technology may in part have been the consequence of the Leven works having been leased for a short time by Daniel Peck, an enterprising native of Cheshire who had been engaged in the salt industry in England,[47] it was also a rational response to the peculiar difficulty of salt making in that location; namely, as Henry Kalmeter noted in 1720, that 'the water here is not so briney as it is at Metthle because a little river of fresh water flows in here'.[48] As this necessitated the use of even greater quantities of coal than was normal to make a 'full' of salt, it was clearly of paramount importance that fuel saving methods be adopted.

For most of Fife's saltmasters however, the search for fuel economies did not become a matter of serious concern until the later eighteenth century when rising demand for coal, including the panwood which had formerly been used almost wholly by the salters, forced up the price and and seriously squeezed profit margins. Even then their response was by no means uniform; while each of the three pans at St Davids in 1788 was 'on Branders',[49] only one of the six going pans at Dysart in 1817 was of this type. Presumably the latter colliery's utter dependence upon the saltwork partially shielded it from commercial pressures which applied elsewhere - without the saltpans to consume the small coal, reported Robert Bald in 1817, 'the Dysart Coal cannot be wrought on account of spontaneous ignition if left below ground'.[50] Even so, the relative lack of interest in the achievement of savings by investing in 'brander' pans was not confined to Dysart. Perhaps this was because, in addition to using fuel more efficiently, the 'brander' pan also produced one fourth more salt than a 'sole' pan; as will be seen in the next section of this paper, increases in salt output for much of the eighteenth century at least were rarely in the producer's best interests.

Fuel efficiency however is only one aspect of the evaporation process; equally important is the matter of the quality of the salt made in the pan. This was determined by several factors, not all of which were in the control of the saltmaster or his employees. Thus in the later eighteenth century, Fife's best salt was said to come from three works, at Drummachie, Largo and St Monance, largely because of the salinity of the sea water they were able to draw on.[51] For the same reason a very high grade of salt was made at Usan, while on Preston Island in 1813 the salt was said to be 'much fairer in colour' than elsewhere, owing to the pan water, which was in this case pumped up from the colliery's 'Engine Pit', containing more salt in a given quantity than Sea Water' and less 'extraneous impurities'.[52]

Conversely, the nearer a saltwork was to a stream or river, the greater could be the problem of making high quality salt. In spite of the efforts made at Leven to reduce costs, the salt made there was said to be 'not so salty by about one third' as at Methil and it tended to become 'moist and damp' when put into a bowl.[53] It was to avoid this problem that David, the 2nd Earl of Wemyss, decided in 1658 to site his pans to the west of Methil harbour 'furthest from the water of Leven', although he was also conscious that there was a quarry nearby which would provide building stone. The strategy was evidently successful in that the salt made there compared well with its counterpart from Leven although, even at Methil, it was noted that 'when an east wind drives up ... it [the water] is not ... so briney as it is otherwise'.[54] The problem was not confined to the seventeenth or eighteenth centuries; in March 1845, the salt agent at St Davids was desperately in need of a shipment of rock salt from Liverpool, owing to 'the great quantity of Snow' which had fallen and weakened the water of the Forth.[55]

Fortunately for the bulk of the saltmasters who had been unable to establish works in the most favoured locations, it was possible to exercise considerable influence over the quality of their product. They could, and often did, ensure that they recruited or trained, and then managed, sufficiently skilled and responsible salt workers. The extent to which the role and status of this occupational sector has been erroneously downgraded by historians will be discussed in Chapter 3. In addition, saltmasters could pay careful attention to evaporation techniques and the acquisition of saltpans of the optimum size. Within certain limitations - the underpan area which could be evenly heated for example - the best salt was made by slow evaporation in large pans. While the dimensions of Fife's saltpans in the later sixteenth century are presently unknown, it seems likely that these were considerably smaller than those which were to become common by the eighteenth century, measuring approximately eighteen feet long by nine feet wide and fifteen or eighteen inches deep. Although it has been suggested that these were introduced to the Forth area in 1810,[56] they had in fact first appeared one and a half centuries beforehand. In 1665, for instance, Lord Wemyss determined that his new pans at Leven should be modelled on what he considered to be an exceptional example and agreed with two masons, John Bruce and John Stine, that they should build him a saltpan and house, 'as large and long as John Wood's pan in West Wemyss or any pan is'.[57] James Meader, a Bo'ness smith, was brought over to construct the pan shell, thereby indicating perhaps that he was familiar with the larger pan size. By 1720 all of the pans at Wemyss were 18 feet long, 9 feet wide and 18 inches deep.[58] Until the later eighteenth century and the survival of inventories which include saltpan measurements, it is impossible to say with any degree of certainty how rapidly the new pan size was adopted. By the end of the century, however, virtually every pan whose dimensions can be ascertained fell into a fairly restricted range, with minimum and maximum lengths of 17 and 21 feet respectively, and widths ranging from 9 to 11½ feet.

In many cases the variations in pan size are hard to explain. Thus in the case of Preston Island saltworks, constructed in the early nineteenth century, there is no obvious reason why one pan should have been 19 feet long, whilst its neighbour was 3 feet shorter.[59] On the other hand those pans which appear to have survived at long-established works, such as Tulliallan, were often smaller than the norm. At the other extreme, the five large pans which were at Culross in 1795 can be accounted for. Although they were then in bad repair, their dimensions, 21 feet long, 11½ feet wide and 20 inches deep, were identical to those at Shields, in north-east England, which in 1748 had been the largest in the country.[60] Clearly the Culross pans, along with a 'House for refining salt', whose internal dimensions were 23 by 21 feet, represented all that was left of the 9th Earl of Dundonald's short-lived scheme, begun around 1786, to make a high quality, large grained, refined salt. So excited had Dundonald been with the possibility of introducing this method to Scotland that he had invited some of the great figures of the Scottish enlightenment, including Professor Joseph Black and Adam Smith, to come and view his 'lately erected' apparatus.[61] For various reasons they each declined his offer. His scheme too foundered; the requisite raw material for refining was Cheshire rock salt, whose entry to Scotland continued to be opposed by the influential salt interest of the Forth.

Dundonald was not altogether alone in his attempts to improve the quality of Scottish marine salt. In May 1758, for instance, Arthur Martin, at Kincaple near St Andrews, was reported to have been preparing to build a second pan of 'an uncommon structure', of wood, which was to be 'a small size at first, for an Experiment ... to be extended if found practicable'.[62] Such cases were rare though. Fife's saltmasters' concern with the quality of their salt was restricted largely to its suitability for storage and transportation without excessive waste. This was less true prior to 1665 which, as will be seen later, marked the beginning of a lengthy period of protection for the industry. As late as 1660, still faced with stern competition in the international marketplace, the 2nd Earl of Wemyss introduced a clause into his salters' contracts whereby he was enabled to deduct a week's wages from anyone who did not deliver clean white salt, which had not been dried according to specific instructions. With the gradual strengthening of their monopoly of the Scottish market thereafter, however, the saltmasters could afford to pay less attention to the quality of their product and, by 1787, it could be openly asserted that they cared little, 'their price being much the same, whatever the quality'.[63] Why should they? In spite of allegations about the dirty grey colour of Scottish marine salt, whose impurities were judged 'unfriendly to the human constitution', merchants appeared satisfied as long as they got 'good weight' while the public at large were 'not judges of the quality of salt', according to one contributor to the *Scots Magazine.*[64] While it was true that the production of a superior salt could give a particular saltwork a slight advantage - St Monance saltworks benefited from this for instance - it was by no means a guarantee of success. William Stewart of Perth, attracted by the price and quality of Largo salt, was advised against purchasing from there on the grounds that the harbour was poor and the scale of operations insufficient to provide a regular supply.[65] In spite of the excellence of its product, Usan saltworks was unable to capture the local market. Even though its price compared favourably with that of its inferior southern rivals, Montrose retailers in 1796 were still supplied with 'considerable Quantities' from the Forth, unwilling it seems to break with that long-established trade.[66]

It was not only differences in salt making techniques which could distinguish one saltwork - indeed one saltpan within a work - from its neighbour. Equally, if not more, striking were the variations in the scale of salt manufacturing operations, which could range from a single pan establishment to a massive battery of ten, twelve or more. Not surprisingly, marked shifts occurred in the relative importance of the various saltworks between the late sixteenth and mid-nineteenth centuries. The most notable were the decline of Culross from its early seventeenth century peak and the addition of a new series of pans in Tayside after 1793.

By the early 1570s salt making had already become firmly established in several locations which were long to be associated with this industry. In 1574 Kircaldy had 28 pans, Dysart 16, Culross 7, West Wemyss 5 and Fordel 4.[67] Not long afterwards there were four pans in the vicinity of Pittenweem. The early lead which Kirkcaldy and its neighbourhood had achieved was almost certainly lost to Culross over the succeeding decades, although the latter's rise was brought to a halt by the devastating storm of 1625. What happened thereafter is unclear. Certainly pan building took place at several locations, including Wemyss and Methil, but the numbers there never rose to the forty which it was said belonged to the Earl of Kincardine.[68] Unfortunately, the lack of technical and output data makes it impossible to ascertain, as seems possible, whether Kincardine's pans, unlike Wemyss's, were the older, smaller, less productive variety. If this is so, and early eighteenth century data points in this direction, it would appear that the ascendancy of the saltworks in the western part of Fife, at Culross, Kincardine, Torry and Tulliallan, was already under serious challenge by the end of the seventeenth century.

By the second decade of the eighteenth century, as Appendix 1 demonstrates, by far the biggest proportion of Fife's recorded salt sales came from Kirkcaldy and its immediate vicinity. Indeed, the region's three largest works, Dysart, Methil and Wemyss, with approximately 24 pans between them, were responsible for an average of 52 per cent of its total sales between 1716 and 1719. The three works in the vicinity of Culross, Craigflower, Torryburn (which as late as 1679 had employed 13 pans) and 'Valefield' sold 15 per cent of Fife's salt. Instead of forty pans at Kincardine and Newpans, only seven were going, selling an annual average of 14,200 bushels of salt between them, equivalent to just under eleven per cent of Fife's total. The smallest works were at Aberdour and Leven Links, where information on the numbers of pans is not available, and Limekilns, which had three pans.

Before the appearance of the Tayside works in the final decade of the eighteenth century, the most striking new development was the construction of nine new pans at St Monance between 1772 and 1775. With the exception of Colonel Wemyss of Wemyss's works this was the largest concentration of pans on the Fife side of the Forth, and contrasted sharply with Leven's two-pan works, or the single pan which was all that survived at Limekilns in 1789.[70] This apart, and excepting too the Tayside works for the moment, the relative positions of most of Fife's saltworks did not shift markedly over the course of the eighteenth century. By 1795-98 Methil and Wemyss were the region's top two producers;

in 1716-19 they had been third and first respectively. Decline continued at Kincardine, which slipped to sixteenth place with a minimal 0.20% of the sales of salt made in Fife and Tayside. Although Limekilns had only slipped one place, to fourteenth, its share of sales had been cut to less than one per cent. Aberdour's saltworks had disappeared, as had those at Kirkcaldy Links, Leven Links, Newpans and Valleyfield. On the other hand there had been some expansion at Inverkeithing and St Davids. At the latter works the number of pans had risen from two to five between 1788 and the early 1790s and, under the close attention of Sir John Henderson of Fordel and his able agent James Pinkerton, had risen to fourth place in the producer's 'league' by 1795-98. (See Appendix 2.)

None of the works established after 1793, either in Fife or Tayside, was very large. Perhaps the biggest was Usan, where three pans appear to have been constructed but, as will be seen later, its productive potential was not fully realised. Although Robert Bald surveyed only two pans when he visited Preston Island in 1813, site evidence indicates that at least one other was added, presumably before 1823 and the subsequent decline of the industry. Thereafter greater uniformity prevailed, with less than a handful of pans in both regions being all that was required to produce what decreasing quantities of salt could be sold. When the report on Wemyss was compiled for the *New Statistical Account,* only one of the pans was still working, making 6,200 bushels of salt per annum valued at £420.[71]

CHAPTER 2

Output, Sales and Markets

By the latter part of the sixteenth century salt manufacturing was by no means new to Fife and had been carried on at various locations along its coast for a considerable time. The early 1570s however appear to mark a critical turning point in the industry's history for exports, which had previously been of little consequence apart from the duration of two 'mini' booms of 1474-90 and 1528-43, experienced a sharp upturn from 1574.[72] While the level fluctuated thereafter and the relative shares of the main markets altered, the export sector continued to be buoyant, certainly until the later 1630s and probably into the early 1660s.[73] There can be little doubt that, as far as the demand side is concerned, much of the increased exporting activity was due to the existence of favourable exogenous factors. The boom of the 1570s coincided with political upheavals and conflict in France and the Netherlands, which interrupted the supply of the much-preferred 'Bay' salt to the seaboard of northern Europe and the Baltic and provided openings for the inferior Scottish product.[74] Similarly, what was probably the best period ever for exports of Scottish salt, the 1620s and 1630s, owed much to the Spanish embargo, begun in 1621, on Dutch trade with Iberia with accompanying steep rises in the price of salt in towns such as Antwerp and Utrecht.[75] Whilst Scottish marine salt was an inadequate substitute for the purposes of the Dutch herring fishery, its applicability for many other processes intensified demand, notably from the Netherlands and the Baltic area, which had previously been supplied, in the main, by the Dutch. The Scots also developed markets in England, Germany and Norway.

The new-found opportunities for the Scottish saltmasters to obtain high, unregulated prices and volume sales made an immediate impact. A conflict between the interests of the producers and the domestic consumers, which was to form a recurrent theme of the story of Scottish saltmaking for the next two and a half centuries, manifested itself early on. During 1574, the Scottish Privy Council heard several complaints of saltmasters ignoring their legal obligations to satisfy the needs of the 'subjects of the realm' before shipping salt overseas and, in addition, charging them 'exhorbitant and unressonabill pryces'.[76] Manufacturing capacity however was rapidly expanded, most spectacularly in and around Culross under the direction of Sir George Bruce,who, between 1575 and 1625, marshalled a veritable empire of forty-four salt pans.[77] The late sixteenth century also saw four new pans constructed at Inverkeithing, by George Gordon of Lawtoun and James Abercrombie of Kerse.[78] Investment in the industry continued into the early decades of the seventeenth century, at Limekilns for example,[79] while in 1628 David, the 2nd Earl of Wemyss, spent 2,500 merks on what he later called his 'old pan'. In 1630 he added a second, this time at a cost of 3,000 merks, which included 'all things, in stonework, timber, iron and lime and workmen'. Great hopes for the venture were entertained and Wemyss recorded in his diary in August 1634:

> The pans with God's help will pay their weekly charge ... with 4,000 merks above the charges. So of this stake 1,000 merks for fresh meat weekly to my house. So rests 3,000m.[80]

Changing political circumstances in mainland Europe, notably the Portuguese Revolt against the Spaniards in 1640, probably burst the Baltic and Low Countries 'bubbles' for the Scots and, by the spring of 1641, the Dutch had managed to regain control of the international salt carrying trade.[81] Nevertheless the Fife saltworks, along with several others on the Forth, did continue to ship salt to the Baltic, although the quantities involved were considerably less than had been the case during the 1630s. However, while the virtual disappearance of the Dutch market itself was undoubtedly a serious blow to the Scots, especially as sales to Norway also appear to have declined, it is important not to overstress the degree of damage incurred by the industry. Whilst sales to the Baltic were below the exceptionally high levels of the 1630s they were still, even in the period 1650-57, almost three times their late sixteenth century average.[82] Furthermore there is much evidence to suggest that exports of Scottish salt to England were increasing.[83]

Certainly in Fife there are no obvious signs of decay: on the contrary, the few indicators we have suggest that saltmasters were anxious to maintain and perhaps even increase output. In the 1640s the presbyteries of Kirkcaldy and Tulliallan made frequent attempts to eliminate Sunday working at the pans, an opportunity for rest which was far from graciously accepted by either masters or saltpan operatives. It was only with the greatest reluctance it seems that the pans at Wemyss were stopped, with the Earl and two neighbouring panmasters, Sir George Hamilton of Blackburn and Sir John Wemyss of Bogie, submitting more than one paper to Kirkcaldy presbytery 'importing the necessitie of the goeing of Saltpans upon the Sabbath'.[84] As late as 1650 John Archibald, an elder of Tulliallan presbytery, was 'exhorted in the name of Chryst to mak more conscience of visiting the saltpans within his bounds' and to deal with those 'who do not put furth their fyre on Saturdayes night'.[85]

While caution should be exercised about generalising from the better documented case of the Wemyss pans, it is instructive to note that, in 1659, the 2nd Earl was confident enough about the future to enter into a five year contract with Robert Hunter, a smith, to maintain the twelve pans at West Wemyss. Even more striking was his decision to extend his saltmaking activities by erecting a range of pans near the new coal port of Methil. By the 15th of July 1665 the smith work on the first pan had been completed and salt making could begin.[86] At first sight the optimism which this activity reveals may appear to have been based more on hope than a careful survey of the trading prospects for Scottish salt. After all, the trend of Baltic sales was downwards and, whilst the Scots saltmasters had apparently benefited from the Cromwellian Union by increasing sales to England, in 1662 the English government had embarked upon a policy of protection for the Tyneside salt industry. The English market was not lost overnight, for recently published figures for Scottish salt imports into London, of 2,828 weys* in 1661-2 and 1,715 weys in 1668-9,[87] point to a marked decline rather than collapse, but by the 1680s, when Scottish salt entering English ports was subject to an impost of 18s.8d. per wey, the trade had virtually ceased.[88]

However, there was another outlet, whose potential has generally been overlooked by historians, and that was the market for salt within Scotland itself. It is true that consumers appear to have preferred French 'bay' salt to the native variety and great quantities were imported into Scotland in the seventeenth century.[89] Despite this, even though the evidence is scanty and altogether unquantifiable, it does appear that there was a fairly significant sale of Scottish salt in the home market, notably to the inhabitants of those areas within the vicinity of the saltworks of Fife and the Lothians, but also further afield. Sometime early in the seventeenth century, one visitor to Fife's southern coast noted that there was made 'an immense quantity of the whitest salt ... which, *in addition to the local consumption* [my emphasis] is sent away for sale to outsiders'.[90] In 1618, at least part of the 90-100 tons being made each week at Culross was being disposed of within Scotland,[91] while Thomas Tucker's 'Report' of 1656 points to the existence of at least a modest inland trade, with salt being sold from the pans 'to cageors and other poor people, who carry the same about in creiles on horseback, or otherwise, up and downe the countrey, for the expence and consumption thereof'.[92]

* a wey of salt was equivalent to 40 bushels, or one ton.

Vast quantities of salt were being imported, however, not only for fish curing purposes, where the Scots variety was generally considered unsuitable, but also for domestic uses. Large tracts of the country depended almost exclusively upon shipments from England, France, Portugal and Spain.[93] It was the prospect of substituting their own product for much of the latter, at least where it was for 'the use of domestik families', which drove the saltmasters in the 1660s to seek the exclusion of foreign salt from Scotland. Although principally inspired by the need to raise revenues, an Act of the Scottish Parliament in 1661 had favoured the Scots producers by levying lower duties on 'inland' salt than on imports, but it was not until 1665 that the protectionist intent of legislation relating to salt was openly declared and the Privy Council imposed an additional £12 Scots on each 'watter boll'* of foreign salt.[94] The exception was salt imported for curing fish and, later, meat which could enter free of duty.

Whilst it would not be appropriate in a regional study such as this to embark upon a detailed survey of the frequent alterations in the 'Salt Laws' which followed, inasmuch as they did have a direct bearing on the industry and its trade in Fife and Tayside two points require to be made at this stage. First, in spite of their growing complexity, from 1665 until they were repealed in 1823, one principle was consistently upheld: duties on Scottish made salt sold within Scotland were to be lower than those payable on imported varieties, except where its use would have been harmful to other national interests, as in curing fish and meat for exportation. Second, in spite of sporadic but occasionally vociferous protests, often from the north east, against a monopoly whereby 'certane noblemen and others in the south' promoted 'their salt pans and trade of whyt salt',[95] and continued flouting of the law by diverting imported salt from its legitimate purposes to the domestic market, sales of salt within Scotland accounted for a growing share of Scottish output. In 1670, if one contemporary estimate can be used as a rough guide, the Scots were consuming around 40 per cent of their native industry's product;[96] by 1706 this may have risen to 69 per cent and, by the second decade of the eighteenth century, to 76 per cent.[97] Thus protected, the Scottish industry was able to survive its losses of sales overseas as well as the potentially devastating impact of competition from Cheshire, which effected the rapid demise of the Tyneside salt-

* Liquid measures were used to determine what quantities of salt were being delivered prior to 1707. A boll of salt was not a uniform measure, but was usually equivalent to either three or four bushels, depending on whether it was a 'small' or 'large' measure. After the Union the bushel, the most commonly used salt measure, was 56 pounds avoirdupois.

masters from the 1730s.[98]

The decision to increase the number of salt pans at Methil in the latter part of the seventeenth century may therefore have been entirely rational. Indeed, the circumstances just outlined go some way towards explaining why, in the face of reduced opportunities for salt sales overseas, there is virtually no evidence of spare productive capacity at home. On the contrary, the late seventeenth and early eighteenth centuries saw the construction of a number of new salt pans in various parts of the country including Fife, at Inverkeithing for example. There, in 1693, the burgh authorities let a piece of ground to David Maither, a smith, 'for his building saltpans and buckit potts'.[99] Around 1705 these were bought by Colonel John Forbes of Pittencreiff, who built a further two pans in the vicinity.[100] While the apparent buoyancy of the period was undoubtedly partly the consequence of the protectionist measures, the outbreak of hostilities with France in 1689, which continued with only a short break until 1713, provided an additional stimulus to the salt-masters of Scotland, along with their counterparts in north-east England and Cheshire.[101] French salt, whether for legal or illegal importation, became scarce and by the late 1690s unprecedentedly large quantities of Scots salt from Kirkcaldy, Methil, Wemyss and elsewhere, were being shipped into Dundee.[102] That Alexander Williamson, a Kirkcaldy merchant, purchased over £596 (Scots) worth of salt from Sir John Wemyss of Bogie in 1702 'for salting and curing of Herrings'[103] is perhaps further testimony of the dearth of the French product, for Scottish salt was generally reserved for salting cod when it was used by fish curers.

The task of plotting the industry's fortunes from the second decade of the eighteenth century until 1798 is greatly facilitated by the survival of the vast bulk of the salt charge vouchers, documents which were completed by salt officers. They recorded the quantities of salt sold from each of the country's works,* thus making it possible to distinguish those in Fife and Tayside from the rest of Scotland. Introduced in 1713, when the new salt duties, levied at rates enshrined in the VIIIth Article of the 1707 Treaty of Union, became payable, the vouchers have the additional merit of listing separately the quantities of Scottish-made salt destined for export and those retained for the home market. Sadly they are not currently available from 1798 until 1823, the period when the Excise Department was responsible for the collection of salt duties.

* Their primary function of course was to collect the salt duties, which were payable when salt was removed from the girnels.

FIGURE 1

Salt Sales from Fife and Tayside Saltworks, 1716-19 – 1795-98

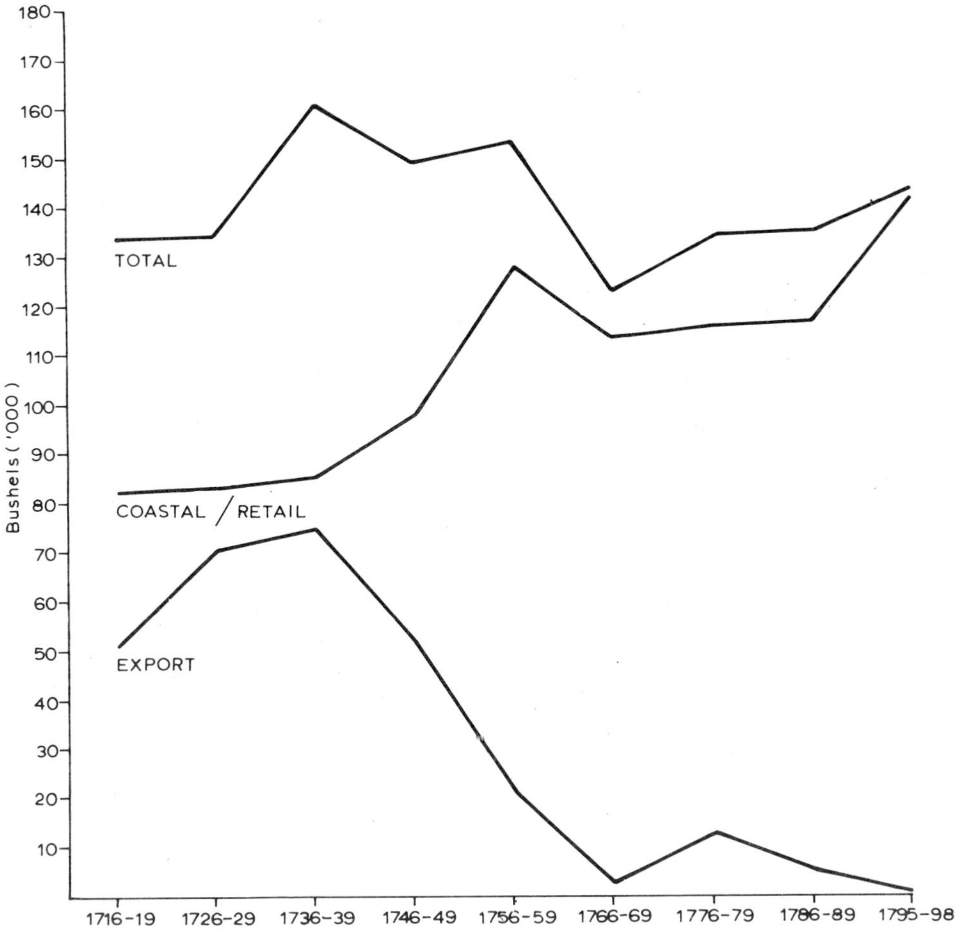

Source: derived from S.R.O. E 536/1-84, Salt Charge Vouchers, 1713-1798.

Note

As production did not begin on Tayside until 1794 figures from the region appear only in the period 1795-98. The graph has been drawn on the basis of 3 year annual averages (at 10 year intervals) with the exception of 1786-89, where data for 1787-88 is missing. The difference between the sum of the coastal and retail sales plus exports and the total sales figures is accounted for by small quantities of salt (never more than 10 per cent, and usually much less) which were used to salt certain types of fish, or lost at sea.

What this evidence demonstrates in the case of Fife is a continuation of trends which had become firmly established in the late seventeenth century (see Figure 1). Exports continued to decline, averaging around 51,000 bushels in the three years from 1716-19,[104] and had all but vanished by the 1790s. The movement was not continuously downwards, however, and during both the 1720s and 1730s shipments abroad increased, largely as a result of events overseas which had the effect of disrupting normal supplies to northern Europe and the Baltic. Such opportunities for volume sales at higher prices than those usually obtainable were eagerly seized and Fife's saltmasters, finding their long established overseas trading habits hard to break, kept a jealous eye on the activities of their neighbours. In January 1744, for example, the Earl of Wemyss urged that a shipload of his salt be dispatched immediately, as he had noted that John Gibson of Durie, at Leven, had stopped 'all sale of salt to land' and 'designs soon to send a loaded ship to Dantzick', an action which pointed to 'ane Information from his brother there, that salt gives a good price'.[105] Such openings, however, became less and less frequent and it seems that the middle of the eighteenth century marked a sharp downward turning point for the export sector, with average sales in the three years from 1756 to 1759 representing a much reduced 42 per cent of the level achieved ten years earlier. The increased availability of cheap rock salt from Cheshire had finally forced the Fifers out of the European market-place. Some shipments of salt continued to be sent out, but this was only possible by selling below cost price, a strategy which was adopted by a number of the Forth-side saltmasters, formed from the early 1770s into the Salt Association, with the aim of maintaining high selling prices within Scotland by restricting supplies and ensuring that stocks were kept at a minimum.[106]

By no means, however, did the loss of export markets herald the end of saltmaking in Fife. Indeed, average annual sales from the county's saltworks were higher in the three years from 1786 to 1789 than they had been between 1716 and 1719. There was a slight fall of under three per cent between 1786-89 and 1795-98, almost wholly due to the establishment of saltmaking capacity within former east coast markets such as Tayside. This will be treated in detail later. That the industry did survive must be considered as a major triumph for Fife's saltmasters, especially as during the second decade of the eighteenth century they had been responsible for around 83 per cent of Scotland's salt exports. How had this loss been overcome? What the aggregate sales figures for Fife conceal is the second of the trends which continued from the seventeenth century, the growing importance of the domestic market. 'Retail' sales, that is purchases of what were generally fairly small quantities of salt, either by persons living within the vicinity of the pans or by 'cadgers' who distributed it further afield, more than tripled between the second and final decades of the eighteenth century. Coastal shipments over the period increased too, but rather more modestly, from an average of 70,300 bushels (53 per cent of the total) in 1716-19 to 86,200 (64 per cent) by 1786-89. Over the same period the combined share of the county's retail and coastal sales rose from 62 per cent (48 and 53 per cent in 1726-29 and 1736-39 respectively) to 86 per cent, and 99 per cent by the end of the century.

A striking feature of this long transformation in the market orientation of Fife's saltmasters was their increased dependence upon sales in Tayside and other locations along or near Scotland's north-eastern coastline. Unlike the fish curers who, like John Richardson of Perth, 'The Great Fishmonger of the Tay', could bring in their preferred Portuguese salt,[107] Tayside's domestic consumers were, legally at least, restricted to the purchase of the Scottish-made variety. For obvious reasons it is impossible to measure the extent of smuggling or indeed how much was diverted from legal purposes into the domestic market without payment of duty. However, it does appear that the considerable consumer resistance to Scottish salt, already identified in the earlier period, continued into the early decades of the eighteenth century; it was alleged that in the 1730s 'foreign Salt was publickly Sold in all the Shops of Montrose and Arbroath', while large stocks of Scottish salt remained unsold.[108] Such blatant examples of illegal dealing however were uncommon and, while clandestine sales undoubtedly continued through the eighteenth century, they were not conducted on an intense enough scale to hinder the growth of a regular legal trade in salt from Fife. The eastern portion of Tayside was supplied by 'cadgers' or salt sellers such as John Marshall, of Menstrie, who would purchase several bushels of salt from the pans in Culross and its vicinity and transport it overland, on horseback, for sale in smaller quantities to the inhabitants of villages such as Doune, Callander and Lochearnhead.[109] But most Scottish salt entering Tayside came by sea.

The trade was dominated by a relatively small number of merchants in the region's larger towns such as, in the 1770s, William Marshall of Dundee, James Low of Montrose and Provost Alexander Simpson of Perth, who placed regular orders with Fife saltworks , normally for shipments of between 1,000 and 2,000 bushels. While some merchants did on occasion play off one saltwork against another in the hope of obtaining a better deal, it was more usual for them to maintain a regular link with a single supplier. There were sound reasons for so doing. 'Transient Customers for Salt' wrote the salt agent at Methil and Wemyss, 'Always pay Ready Money'.[110] Only their 'best Customers who deal with no body Else' obtained salt at prime cost with two or perhaps three months credit. Regular dealing was also more likely to ensure that orders were fulfilled during periods of scarcity; thus, in September 1771, when the above-mentioned agent complained that he had been 'these past several weeks nearly pestered out of my Senses for salt' and could make no promises about supplies in the foreseeable future, he was careful to ensure that his 'good friend Provost Simpson of Perth', an important customer, obtained what he needed from Bo'ness saltwork. Similarly merchants with whom the saltmasters or their selling agents were familiar could expect to be supplied with 'old' salt, which they much preferred as it had been allowed to drain for some three months and was therefore less liable to liquefy, and so reduce in quantity, during the sea voyage north. In December 1775, George Jack in Perth was assured that he was being sent salt which was 'good and dry' and would unload well 'if the Vessel gets Safe up [the Tay] without being Obstructed by frost'.[111] The problem of wastage however was not easily overcome; even in 1847 when, like most other surviving Scottish saltworks, the salters at St Davids, Inverkeithing, were mixing Liverpool rock salt with sea water to produce salt 'of an excellent quality and fit for immediate use', James Miller of Perth was asked if shipment could be delayed for a week as the salt would then be able 'to bear the fatigue of shipping'. When the cargo was eventually dispatched in May, five hundredweight was added to compensate for the anticipated weight losses en route.[112]

While the development of a close relationship with a supplier was clearly beneficial for an importer of salt, for many of Fife's saltmasters such arrangements were of crucial importance. This was especially so in the cases of those who, prior to its post-mid-century collapse, had been heavily engaged in the export trade. It is true that by the second decade of the eighteenth century the largest proportion of the county's saltworks had ceased to be heavily dependent upon overseas sales. Between 1716 and 1719, for example, less than ten per cent of the salt made at Colonel John Erskine's works at Newpans was exported, while not a single bushel went overseas from Lord Colville's Craigflower pans. To some extent saltworks such as these, in south-west Fife, were favoured by their geographical location, being within range of land transportation to the rising markets of central Scotland. Indeed, by the later eighteenth century, salt from Culross and Torryburn was being sent as far west as Dumbarton and Glasgow, in addition to Alloa, Clackmannan, Kinross and Stirling.[113] For those saltworks in the neighbourhood of Kirkcaldy, however, very different conditions prevailed. Between them the five works of Dysart, Kirkcaldy, Kirkcaldy Links, Methil and Wemyss were responsible for some 83 per cent of Fife's salt exports. Of these Methil and Wemyss would feel most heavily the consequences of reduced opportunities abroad, for in the period 1716-19 exports had accounted for a substantial 64 per cent of their sales. For no other saltmaster (both sets of pans were owned by the Earls of Wemyss) were overseas sales of such importance. Even as late as 1786-89, 34 per cent of their sales went overseas. Nonetheless, along with their neighbours, they had a substantial proportion of their sales to direct elsewhere. Even though landsales increased markedly, as they accounted for only 29 per cent of the county's total sales by 1795-98, it is clear that the industry's survival depended upon its ability to maintain and increase coastal shipments. As the south-east of Scotland was supplied from the saltworks on the south bank of the Forth,[114] it was upon the market potential of Tayside and the north-east that Fife's saltmasters increasingly had to depend.

Unfortunately, the salt charge vouchers list only general sales headings such as 'retail' and 'coastal' and thus provide no clue to the specific destination of the salt sold from the pans. In order to establish where it was being sent, it is necessary to rely on the considerably more arbitrary evidence of a very limited number of surviving saltwork sales and letter books. The picture which these present is fairly clear. Of coastal shipments from Methil and Wemyss for instance, which had risen to 53 per cent of their sales by 1786-89, over 80 per cent were going to Tayside and the north-east in the 1770s.[115] On occasion, as in the twelve months from January 1771, Tayside formed the largest single coastwise market for these two works, with over 42 per cent of their coastal sales being sent to Dundee and Perth. The figure rises to 44 per cent if a single shipment of 760 bushels to Newburgh, on the south bank of the Tay, is included. In the same year, 38 per cent of the coastal shipments from Methil and Wemyss went to ports further north, such as Aberdeen and Nairn, with the rest going to merchants in Leith and Edinburgh. It is little wonder then that Provost Simpson of Perth received such favoured treatment in 1771; it was estimated that he was then buying one quarter of the two works' annual output.[116]

While their relative positions shifted - in 1773, for example, the north-eastern merchants took 40 per cent as opposed to Tayside's 38 per cent of Methil and Wemyss's coastal sales - the market to the north of Fife had clearly become indispensable. Its importance may have been unevenly spread; Sir John Anstruther's saltwork at St Monance for instance was able to dispose of more salt to the fishing fleet than was the case with most of Scotland's other saltworks,[117] while in the mid-1790s around half of that work's output was purchased by John Turpie, a Glasgow merchant,[118] but even at St Monance, as David Loch reported in 1778, 'many north country vessels' arrived to take on coal and salt.[119] The fact is that few of Fife's saltmasters could have remained in business from the mid-eighteenth century onwards without their coastwise shipments. In the late 1780s only two works, Craigflower, near Culross, and Limekilns, relied solely on landsales.

The home market then compensated for Fife's export losses. In spite of this however, indeed partly because of it, the industry was in an extremely vulnerable situation. That the saltmasters had been able to respond to the short-lived spurts in overseas demand during the 1720s and 1730s by increasing output suggests that there may have been a surplus of salt making capacity in the more subdued trading conditions which prevailed for most of the eighteenth century. The problem was partly eased by the exhaustion of accessible coal reserves in some localities, with the result that saltmaking ceased.[120] However, as Appendices 1 and 2 demonstrate, only five saltworks which had been going in 1716 had ceased to exist by the end of the century. On the other hand some works had been greatly reduced. All that was left of the formerly impressive works at Tulliallan by 1795 was one working pan, with three others in various states of disrepair. Of one there was 'nothing but the walls'.[121] Nevertheless, a major problem remained, and it seems quite reasonable to claim that the success of the pans which continued to operate depended very heavily upon the maintenance of a delicately balanced equilibrium between supply and a geographically restricted demand, whose expansion was almost wholly determined by the rate of population growth in Fife, Tayside and the rest of north-east Scotland.

Occasionally potential threats to the industry's stability did appear as in January 1758, when Arthur Martin was reported to be erecting a salt pan and girnel at Kincaple, in the parish of St Andrews.[122] However, unforeseen difficulties, not least that of getting coal to the saltwork, caused Martin's hopes to be dashed and, while some small quantities of salt were made there, the salt officers in attendance at the pan were withdrawn early in 1766, on the grounds that there had been 'no receipt upon Salt for a considerable time past', a clear indication that production had ceased.[123] Equilibrium was thus maintained. A far greater threat, and indeed one which did materialise, was presented by the re-establishment of saltworks on the shore between Pittenweem and

St Monance. In the latter part of 1770 demand for salt became intense and prices began to move upwards. By March 1771 it was claimed that 'there is not at present any salt to be had in all this Frith', while at Methil the girnels were completely empty, 'a Circumstance never known by any about the Works'.[124] Undoubtedly it was this healthy prospect which in large part inspired Sir John Anstruther and his partners in the Newark Coal and Salt Work Company to construct what was to become one of Fife's largest works. The first pan was completed in July 1772, with another five being commissioned by the end of the year. Two others followed in 1773, while a ninth was completed in June 1774. The location of these and other pans, as well as the dates after which salt making could legally begin, can often be traced through the Customs and Salt office letter books. Officers were required to be in attendance at the pans from the moment salt began to be made, and therefore close attention was paid to building activity in the vicinity of saltworks. The completion of a new pan had to be reported to the nearest salt officers. Thus on the 7th July 1772 Philip Brown, agent for the Newark Coal and Salt Work Company, 'Proprietors of Salt in the Parish of St Monance and Shire of Fife' did:

> ... hereby Enter with Andrew Johnston Esq. Collector of Salt Duty at Anstruther One Salt Pann, Bounded on the East by One Salt pan Now Erecting, On the South by the full Sea (a foot road intervening), On the West by Three Pans Erecting and upon the North With the Salt Girnel and Corn and pasture field, Together with the said Girnel itself.

Similarly, some years earlier, at Kincaple, John Hope had entered, on behalf of Arthur Martin:

> One salt pan bounded with the Sea upon the North, the salt girnel and King's High Way upon the South, upon the East with Corn Fields and the West with an Earthern trow* for Conveyance of the Water into the Pans....[125]

Such activity in the early 1770s was not confined to Fife. Throughout the salt producing areas of Scotland formerly 'silent' pans were rehabilitated and others added.[126] Consequently when the boom broke, its peak appearing to have been passed by the end of 1772, the industry was left with a large stock of surplus salt making capacity. In order to avoid what would have been suicidal, cut-throat price competition the Salt Association was formed, probably in 1773. As it operated on a national level and its membership comprised most of the leading salt-

* a trough.

masters from both the north and south sides of the river Forth, lengthy discussion of its affairs cannot take place here. However, it is important to recognise that, for Fife's saltmasters, the existence and successful operation of what was referred to in correspondence as 'the Society' was crucial. During the following twenty years they often played prominent roles in its organisation, as in the case of William Wilson, sometime secretary, who worked from Wemyss Castle, and Sir John Henderson of Fordel, President in the early 1790s. With such a small proportion of their output being sold in Fife and, at St Monance, the addition of as much as 40,000 bushels of productive capacity, equivalent to around 32 per cent of the region's annual recorded sales between 1766 and 1769, it was imperative that they eliminated competition in the 'seasale' or coastal market. For some it was not solely the survival of their saltworks which was at stake; without the outlet for the sale of 'small' coal which saltmaking provided, several of the Forth-side collieries would have been in grave financial difficulties while those at Dysart, Wemyss and St Monance were said in 1788 to 'depend entirely upon their Salt'.[127]

By restricting their output to an agreed quantity determined by the number of pans at each work, imposing a tax on overproduction, fixing selling prices and disposing of surplus salt 'to Fisherys, England, or Exportation, but not to Scotland',[128] most saltmasters managed to remain in business. Of those works which had been going in 1776, only two, at Drummachie and Largo, were no longer in existence two decades later. While few if any fortunes were made from salt, the Association does appear to have managed to match salt supply broadly with available demand. In 1793, however, two events combined to upset the balance by threatening the monopoly of Scottish salt sales which the Fife saltmasters had in Tayside and the north-east. First, in response to complaints about shortages and excessively high prices, the duty of 3s.8d. per ton of coal sold north of the Red Head, at the southern end of Lunan Bay, was removed,[129] and secondly, largely owing to war-induced shortages, the price of salt rose to an unprecedentedly high level.[130] Consequently, perhaps for the first time ever, there was a fair prospect of making salt economically at locations along the non-coal-bearing coastline north of the River Tay. Such new-found opportunities were quickly seized. By May 1794 James Dickson, junior, of Montrose had purchased a house from the Greenland Whale Fishing Company, which he intended to convert into a saltwork.[131] In June, David Scott of Dunninald, M.P. had begun building two pans at Usan, on his brother's estate.[132] These he hoped would eventually make 3,000 bolls (around 9,000 bushels) of salt per annum and supply one-third of the inhabitants of Angus with their requirements of 'that important and necessary article'.[133] By the end of 1794 salt production had begun at both works and early in 1795 the Dundee Glasswork Company had also entered the salt making business, in the vicinity of Carolina Port.[134]

The available evidence suggests that these schemes may have been somewhat less successful than their promoters had anticipated. Certainly in February 1796 the Customs Office in Montrose described the duties collected at Usan as 'tryfling'[135] and, even with the addition of a third pan, recorded annual sales had not exceeded 3,700 bushels by 1798. Further, in spite of the recognition that good quality, competitively priced salt was made at both Montrose and Usan, local retailers continued to import salt from Inverkeithing, Kirkcaldy and elsewhere.[136] In the two years from July 1794 just over 14,334 bushels were brought into the Montrose district for domestic consumption, while the two local works were forced to dispose of 3,950 bushels coastwise. Tayside's salt retailers were evidently unwilling to break with suppliers whose shipments, although dear, had been regular.

Nevertheless, between them, the saltworks at Dundee, Montrose and Usan did sell an average of 12,200 bushels in the three years from April 1795 and, accounting for between eight and nine per cent of the two region's aggregate sales, clearly cut into the market for Fife's salt. Furthermore they were able to sell most of their salt locally. In the three years from 1795 just over 70 per cent of their sales were within Tayside, 23 per cent was sent by sea to other parts of Scotland, and a marginal 7 per cent was either exported or used by the fishing trade. In the same period St Monance saltworks was only able to dispose of 13 per cent of its salt locally. Whilst David Scott's ambitious hopes may not have been realised, his works at Usan did rank a respectable eleventh of sixteen amongst those in Fife and Tayside, between 1795 and 1798. James Dickson's Montrose works did even better, producing six per cent of the two region's salt and, ranking eighth in the sales league, stood one place above the long-established Leven saltpans then being operated by James Christie. For Fife's saltmasters the problem was not restricted to competition from Tayside. Saltworks were opened further north too, at Nigg, Peterhead and Portsoy.[137] Fortunately for the Fifers, the damage which this activity might have done was limited in its extent. Fife's salt sales did fall by just under three per cent between 1786-89 and 1795-98, compared to a Scottish increase of 14 per cent. On this occasion, retail and coastal sales were insufficient to compensate for a further, and almost final, collapse of exports. However, while it was undoubtedly the presence of the new saltworks in Tayside and the north-east which produced this net loss, coastal shipments as well as sales within Fife did increase, largely as a result of the continuation of the war with France. This ensured that demand for the region's salt remained vigorous and, after a short period of uncertainty, confidence was restored. As early as 1795 the minister of Kirkcaldy parish was able to report that the pans there had been restarted while, in July 1796, some prospective lessees of Dysart saltworks informed Sir James St Clair Erskine, the owner, that they wished to put the existing plant 'in better order than at present' and build an additional, eighth pan.[138]

The period from 1798 until 1823 began hesitantly for the Scottish salt industry, with sales generally remaining below the levels achieved in the preceding three years until 1806.[139] Recovery followed thereafter and for sixteen years sales reached previously unrecorded levels. On the other hand with the disappearance of the Salt Association and thus the absence of self-regulated marketplace discipline, it appears that increases in the volume of output were accompanied by low profit margins. Although it is not possible to obtain separate figures for Fife and Tayside, there is no evidence to suggest that the experience of the two regions diverged markedly from that of the industry as a whole. One clear sign of business optimism was the construction of new saltworks. At least two were opened in Fife, both in the early nineteenth century. Unfortunately there is virtually no documentary trace of the work built at Crail.[140] About its near contemporary, on Preston Island, off Low Valleyfield, more can be said. When he visited the colliery and saltwork there in 1813 Robert Bald, the renowned Scottish coal mining 'viewer', found two saltpans in operation, producing 130 bushels of salt per week between them, though probably making little profit.[141]

Tayside's fourth saltwork, at Arbroath, was opened by James Butchart in 1814.[142] Although this may have been the last to be built within Tayside, it did not quite mark the end of the challenge to Fife's saltmasters. This came from Brora in Sutherland where renewed attempts - the first had been made in the late sixteenth century[143] - were being made to mine coal and manufacture salt. During the construction of the pans in 1814, the Marchioness of Stafford reported to her husband that the 'Fife people' were 'more uneasy about our Salt than any Thing else' as one-third of their business was 'the supply north of Peterhead'.[144] Four pans were eventually built, whose salt satisfied a good part of the demand on the Moray Firth, at two-thirds of the previous retail price.[145] How serious the impact of this was for Fife's saltmasters is difficult to judge. Clearly the shipment of large quantities of salt from Brora, almost 11,900 bushels in 1818 for instance, was a matter of considerable concern. However as the effect of this was likely to have been spread fairly evenly between Fife's saltworks, whose combined sales were 138,300 bushels in 1797-98, the impact of the development at Brora was not sufficiently serious to have justified earlier fears. On the other hand, the addition of yet another rival in a formerly secure market made life even more uncomfortable for the southern saltmasters.

Within a decade though the final blow had been struck. On the 5th day of January 1823 all duties on salt made in Scotland or entering the country from England were swept away. Exposed at last to the cold winds of the free market, the makers of the expensive Scottish product were unable to stem the inflow of the purer, highly concentrated and considerably cheaper rock salt from Cheshire.[146] With the exception of Arbroath, all of Tayside's surviving works closed within a few years of the repeal of the salt duties, unable to bear the cost of fuel and its transportation which their situation necessitated. In Fife the industry was reduced rather more slowly, but the rate of decline was by no means uniform. At Dysart, the proprietor's royalties from his lessee's saltwork profits dropped sharply, from over £151 and £122 in 1824 and 1825 respectively, to £30.6s.8d. in the twelve months to May 1826.[147] Similarly at Methil and Wemyss output fell steeply. When the *New Statistical Account* was written, salt making had disappeared from the former locality while at Wemyss only one pan remained in operation.[148] The experience of St Davids saltworks was somewhat different. Profits did fall, from £412 and £951 in 1823 and 1824 to an average of £203 per annum between 1828-30 (inclusive) and £127 per annum ten years later.[149] Output however was reported to have been 30,000 bushels in 1836,[150] which represented a considerable increase over pre-1823 sales figures. It appears that saltmaking continued here, as at a handful of other locations, including Arbroath, by the importation and refining of rock salt shipped from Liverpool. The high quality salt which resulted from the process of boiling rock salt in sea water was saleable both locally in Fife and through the traditional outlets such as Perth and Montrose. Indeed, on occasion shortages appeared; in November 1846, the salt agent at St Davids informed James Miller in Perth that 'the demand for salt is so great I am doubtful whether I will be able to load a vessel up'.[151] Furthermore, the previously unsaleable 'bittern', a magnesia-based compound which drained from the salt, found a market in the growing chemical industry; in the 1840s, regular loads of six or eight barrels were sent from St Davids to Thomas Astley's chemical works in Bonnington.[152]

The trend however was set firmly in a downward direction. Around 1850 the works at St Davids closed,[153] even though in 1836 it was claimed they were constructed 'upon the best principles', while James Townsend Oswald's Kirkcaldy pans, 'capable at present of making Three Tons of Salt weekly', were without a lessee in 1852.[154] A few works did manage to struggle on into the second half of the nineteenth century. Slater's *Directory of Scotland* included five salt manufacturers from Fife in 1873, at Inverkeithing, Kirkcaldy, Leven, Limekilns and Wemyss, as well as one from Tayside, the Arbroath partnership of John and Alexander Ferguson while, by 1900, a new saltwork, at Kinghorn, had come into existence. This was exceptional, however, and one by one the last representatives of the industry disappeared. The closure of the pans at Limekilns in 1946[155] probably marked the end of Fife's direct connection with salt making.

CHAPTER 3

Salter Serfs

Traditionally salters have been seen as members of an occupational group whose circumstances were barely distinguishable from those of the colliers. With them they shared the burden, after 1606, of increasingly hostile parliamentary legislation, the courts' interpretation of which merely served to undermine further their already lowly occupational status.[156] The salters, like their collier brethren, were bound to their place of work for life. It has been argued that this was possible partly because the job of making salt, like mining coal, was unskilled and 'therefore easier to organise along serf lines'.[157] Indeed the two industries were closely related, by common ownership or joint leasing of integrated colliery-saltwork enterprises, the rationale for which lay largely in the fuel requirements of salt making - small coal. Further, as the salt pans and the coal pits which supplied them were rarely more than two or three miles apart, the colliers and salters invariably lived within the same community. Yet there were differences between the two groups of workers, the colliers working underground whilst the salters, of whom there were generally far fewer belonging to any combined colliery and saltwork, remained on the surface, employed in individual panhouses performing a task which demanded very much longer periods of constant attendance than was the case in mining. Neither, it seems, did the colliers and salters live together in a homogeneous group. On the shore at Culross for example it was possible in 1795 to distinguish the colliers' housing from the two salters' rows, organised in two ranges, one 118 feet long, 15 feet deep and 8 feet high, the other 33 feet long, 20 feet deep and 18 feet high. At Kincardine too, where both industries were carried on, there were four two-storey salters' houses, distinct from those used by the colliers. Little attention however has been paid to the salters specifically, and those judgements which have been passed paint a bleak view of their situation. The historian of Culross and Tulliallan contrasted the 'wild existence' of the unlettered 'pariahs' on Preston Island with 'the more civilised inhabitants of the adjoining shore',[158] while, according to the Clows, in the most extensive treatment of the Scottish salt industry which is currently available, the 'conditions of labour in the Scottish salt-works were ... ghastly' not least because the salters' wages were paid 'in the commodity they produced'.[159]

While a limited regional study such as this cannot hope to do justice to the major historiographical task of rescuing the Scottish salter from the anonymity into which the shadow of numerous studies of coal mining labour has thrown him, there is enough evidence from Fife to suggest that his lowly legal status and relative immobility should not be allowed to conceal the very considerable influence he could exert within the workplace, both on his own and in combination with his fellow salters. The spectacular increase in demand for salt in the late sixteenth and early seventeenth centuries presented the saltmasters with novel problems of both labour recruitment and control. A partial answer was to allow coal and saltmasters to 'tak and apprehend all maisteress vagabondis and sturdie beggaris and put thame to woirk' and later, in 1641, to restrict their movement by placing a legal maximum on the bounty or fee which could be paid to a worker coming to a new work.[160] At the same time, as has been seen earlier, attempts were made to enforce a more regular working pattern in both coal mines and salt works. These measures were only of limited assistance to Fife's saltmasters; at no time did they manage to achieve complete mastery over their salt workers. There were three principal reasons for this, first the requirements of the job of making salt, second, the nature of the contractual arrangements which bound master and man together and third, the physical environment in which the work was done.

Unfortunately, those writers who have described salt making as 'unskilled' have not made it clear what they meant to convey by the use of that term. It would be wrong, however, to assume that the job of making common salt was one to which any individual could be directed, without a considerable period of training. If little skill was involved, it is difficult to explain why saltmasters went to such considerable lengths, when they required additional salters, to recruit people with previous experience in the industry. During the buoyant years of the later 1720s, for example, the salt agent at Bogie not only searched for workers in Fife but also scoured the south bank of the Forth, from where he brought at least two salters. Thus in June 1727 over £8 (Scots) was spent on shipping over 'Mark Muir, Salter his familie and plenishing',* an expense to which 10s. for ale was added a week later on the occasion of the 'salters aggreement and ... ther welcom over from Lothian'. By October, Thomas Curzon, from the same area, had also been enticed to Kirkcaldy.[161]

The attributes required of a salter, as well as his role and status within the salt making community, are best understood through a brief examination of the organisational structure of a saltwork. Most owners or

* furniture and goods

lessees of saltpans, the 'saltmaster' or 'panmaster', employed a factor, 'salt agent' or 'saltgrieve', at a fixed salary, whose function was to manage and co-ordinate the complete saltwork, regardless of the number of individual pans. Beneath him and central to the whole operation, was the 'master' salter, usually described simply as a 'salter' although the more precise account books rightly record the prefix 'master'. The master was subordinate to the saltgrieve but his responsibility was to operate a saltpan and make salt. Either individually or in combination with others, the master salter agreed with the saltmaster or his agent to make a certain quantity of salt for a fixed amount of coal. The ratios could alter, when a different type of coal was used for example, but from the sixteenth to the nineteenth century the basis of the arrangement remained the same. As good an example as any is the agreement reached between the 2nd Earl of Wemyss and his master salters in 1659. Wemyss recorded that:

> For every twelve loads or dozens of coals here set down the salters have obliged themselves under their hand writing ... to pay to me and my servants two bolls and two pecks of salt for each dozen of coals I give them....[162]

For his part Wemyss was to pay them 3s.4d. (Scots) for each boll of salt they made and delivered. A salter who failed to deliver the agreed quantity would be penalised by a fine of 30s. for each boll that he (and very occasionally, she) was short. In their details, contracts varied over time and between saltworks. In 1660, as was mentioned earlier, Wemyss added a new clause by which his salters were to 'give sufficient white salt well dried and not to sweep [ashes from under] the pans when they draw the salt, under pain of losing that week's wages'. In spite of the apparent harshness of this rule, he was later gratified to reflect that his arrangements 'make me in as good condition with my salters as any in Fife are', not least because, whereas he contracted to pay either money or meal to his salters when the pans were stopped for any good reason, those working for Lord Sinclair at Dysart had 'nothing but his goodwill'.

There were at least two good reasons why this category of saltwork operative earned the title 'master' and the craft-like status thereby implied. As it appears that most salt workers were recruited from within the industry - with fathers training their sons, who then became 'master salters' in their own right - the documentary evidence concerning the skills which a salter might be expected to attain is extremely thin. However it is not altogether absent, and the procedure by which a new entrant to the industry learned his skills can be traced. The case of

Walter Smith, from 'Auchtergevan in Perthshire' is especially instructive.[163] In February 1752, Smith signed a lengthy agreement with David Beatson, factor to Lieutenant-General James St Clair of Dysart, in which Smith promised to remain at Dysart for 14 years, during which time he would be instructed in the 'Art of Saltmaking'. Although no other periods were laid down, the method by which he was to be trained was. Instruction was to be provided by 'one or other' of the salters at Dysart, who would 'Conceall no point nor practique' from Smith. Until he was 'able to work half ane pan', payment, in the form of clothes, maintenance and two pecks of oatmeal each week, was to be made by Beatson. Thereafter, 'being ane able and Sufficient Salters Servant', Smith was to be paid 'the same fees as any other Salters Servant at Dysart', until he was 'Capable ... to manage and work ane Saltpan'. At that stage he would become the master of his own pan, on the same terms as the other master salters, that is 'Conform to the Quantity of Salt to be made by him after discounting the pryce and value of the Panwood to be made use of by him'.

The justification for such formal training is not hard to find. As was widely recognised, the production of good salt, without burning the pans, was a 'nice operation'. The more rapid the boiling process, the less satisfactory was the salt, as it was more likely to liquefy. At least equally important was the damage which overheating, uneven firing or ignoring the need for regular 'paidling', could do to the pans themselves, the cost of repairs to which was the responsibility of the saltmaster and his grieve. Thus, while the 2nd Earl of Wemyss was anxious to obtain dry white salt from his salters, he was even more concerned about the care of his relatively expensive capital stock; any salter who 'burn and spoil his pan knowingly ... is to be punished in body and purse', he noted in 1657. Evidence from elsewhere in Scotland suggests that saltmasters were prepared to let unsatisfactory salters find alternative employment.[164]

It was not only skill and a considerable element of responsibility which earned some saltworkers their 'master' status. By no means did they occupy the lowest rungs of the saltwork hierarchy. Such places were filled by the 'bearers', sometimes female, who carried coal from the pits to the pans, and old women and others who were employed in washing salt 'pocks' and mats and other menial tasks. Directly below the 'master' came his 'servant' or assistant, often also termed 'salter', without whose labour the master was unable to operate the pan. During May 1723 for example John Foord, a master salter at Bogie, was unable to make salt for a week as he was 'wanting a servant'. On more than one occasion between November 1725 and May of the following year John Hound's lack of output was explained by his 'wanting a man', although an entry for June in the account book, of 15s. (Scots) 'To My Lord Weems officers for bringing a Salter Boy', indicates that his problem was soon overcome. The time consuming and physically taxing job of making a 'full' of salt could not be carried on by one person. Assistance was required in keeping the fires going, constantly and evenly, in clearing out the ashes, raising water into the pans, stirring the brine and drawing the salt, tasks which had to be carried out during the day and night. Although the family working unit appears to have been common on the south side of the Forth, notably in the Prestonpans area,[165] the evidence currently available for the Fife side suggests that salters' servants were more likely to have been unrelated to the master and rather employed by them. Boys too - apart from the adult assistants - were employed. Thus each saltpan was operated by three workers, in a well defined triple-tier system. This was clearly the case at Methil in 1772 when the winter allowances were paid. Eight of the nine master salters - the exception was Robert Baird 'whose Pan was sett* in Octr last' - were given several free loads of pan coals, the same number of salters' servants 3s. each and 'Each of the boys who are paid weekly by their Masters - two shills and Sixpence'.[166] Although the saltmaster or his grieve usually recruited the salters' servants and boys, and paid their annual 'binding' money, it was the master salter who was responsible for their weekly payment and conduct. William Foord, a master salter at Bogie in 1732, had his 'plenishing' confiscated by the grieve on the grounds that his salt arrears were extraordinarily large, even though Foord claimed that this was caused by the 'unfaithfulness of my Servant' who 'manie a night when it was his turn to work ... hurt the pan [and] neglected his work and imployed him self in the breaking oppen Sr Jeames girnels', presumably with the intention of stealing salt.[167]

Effectively then, this was a system of sub-contracting; the saltmaster or proprietor was only indirectly involved in the production process, whilst the master salter became an employer of labour, albeit on a small scale. This contractual arrangement was not unique to saltmaking in Fife, or indeed Scotland, but was also found in Cheshire, the heartland of British salt production, where it was 'the original basis of employment' and survived, as it did north of the border, well into the nineteenth century.[168] In Fife, however, as elsewhere in the main Scottish salt producing regions - the exceptions being Galloway and Orkney - the

* leased, or in this case occupied by Baird.

basic contractual arrangement was underpinned by the labour legislation of 1606 and 1641 as well as by the draconian powers held by the Scottish landowning class to control those over whom they had any authority. How far did these factors affect the salters and to what extent was their situation made worse by the conditions of serfdom to which they were legally subject until the system was finally ended in 1799?

There is a deal of evidence, from Fife as elsewhere in the Scottish salt-making localities, of harsh and sometimes brutal treatment being meted out to salters. As has already been seen, wilful damage to a pan could result in corporal punishment being inflicted upon the offender. In 1728, the 4th Earl of Wemyss, on hearing that 'the salters have been so insolent as to deforce William Thomson, the officer, from apprehending Rob. Mitchell', presumably for 'running' or selling salt without payment of duty, ordered that the culprits be incarcerated in Kinghorn prison, while 'young Miller, who was the ringleader' was to be put 'into the pitt' (a type of dungeon).[169] Of course, arbitrary and severe treatment of those who strayed outwith the bounds of acceptable behaviour, within the confines of the brutally paternalistic Scottish burgh of barony, was not reserved solely for either salters or colliers. Masters of other types of early manufacturing industry, seen in the case of cloth making at New Mills, used imprisonment as a means of inculcating order and obedience.[170] And the wild and bedraggled appearance of the salters may have been far less a consequence of their serfdom than of the nature of the job itself, which was carried on in cramped, dark, hot, smoke and steam-filled panhouses. Similar sights and conditions were to concern late nineteenth century factory inspectors of Cheshire's salt 'wych-houses', where 'serfdom' was unknown.[171]

Perhaps the major burden under which the Scottish collier-serfs laboured was that they could not move freely from colliery to colliery; they were, in effect, bound for life.[172] Coalmasters were frequently involved in searches for 'deserters' or in litigation with their fellow-masters about colliers whose ownership was in dispute. Examples can be drawn from Fife's saltmasters and salters to demonstrate that this was common ground shared by the two industries. In November 1636, the Laird of Tulliallan's brother was accused of taking John Wilson, a salter, from Elphinstone pans, and in order to get him into a boat to cross the Forth, was said with his accomplices, to have put 'violent hands on said John ... and strake him' and then detained him at Tulliallan pans.[173] Life bonds were entered into, by James Alexander and Peter Pugston for example, who in 1757 obliged themselves and their heirs to serve Sir James Wemyss of Bogie 'in the Station of ... Salter, all the Days of our

Lives'.[174] Like the colliers, salters were sometimes considered to be private property, to be leased or sold by their owner. Sir Robert Henderson of Fordel let 'Three Salt panns Salt Girnells Salters and Salters houses ... lying at Inverkeithing' to John Wilkie for 300 merks per annum from 1746;[175] in 1705, the 'Laird of Bogie' invoked the full force the law to arrest and have returned Patrick Davidson, a salter, who had left him to work at Maitland pans, near Bo'ness.[176]

Yet proportionately such cases appear to have been far less common amongst salters than coal miners. Leases which included salters along with the stock of a saltwork were rare and searches for absconders, unusual in the 1600s, were even less so in the following century. A striking feature of the few lists of salters' names which have survived is the frequency with which the same names recur, often with the suffix 'y^{r}', or 'younger' added, indicating a notable degree of continuity of employment within the industry. No case of disputed ownership of a salter appears to have reached the Court of Session during the eighteenth century. If anywhere near accurately drawn, this picture of an industry which was relatively little troubled by the problems of labour supply and instability in the later seventeenth centuries contrasts sharply with coal mining, and clearly requires explanation.

For Fife's saltmasters, the potential difficulty of recruiting new salters hardly arose, owing largely to the prevailing market circumstances. From the mid-seventeenth century, apart from generally short-lived periods of unusual buoyancy, the total size of the available market probably experienced little significant alteration. As long as most salters who had been recruited during the rapid expansionary phase of the late sixteenth and early seventeenth centuries remained within the industry and merely replenished their own stock by natural means, few shortages occurred. Certainly there appears to have been no labour shortage in the 1660s, as not only was the saltmasters' plea for protection in 1665 based partly on the grounds that failure to restrict foreign imports would be 'the undoing of many thousands of poor people', but so too was their insistence that the measure be retained in spite of popular opposition to it. Favourable conditions for sales in the later seventeenth and very early eighteenth centuries may have caused the masters to tighten their grip on existing salters, and perhaps even engage in a modest search for new recruits but, as the century advanced, such occasional forays into the labour market as did occur can invariably be linked with short-run upturns in the industry's fortunes, during the 1720s for example, or with the specific needs of a particular saltwork. To some extent the same circumstances explain why so few of Fife's salters deserted their masters. Having adequate supplies of labour it was only in exceptional circumstances that saltmasters were prepared to tempt salters to leave their lawful masters by paying a sufficiently large binding fee, as well as risking the inconvenience and cost of legal proceedings which might follow. As new saltworks opened only occasionally, and those which existed were well-known and limited in number, opportunities of finding suitable work, especially for the 'master' salter, were extremely limited. After all there were, in 1787 for instance, only 118 pans (and thus jobs for 'master' salters) in the whole of Scotland, 62 of which were in Fife. Physically too it was far more difficult to conceal a salter than a coal hewer, who worked deep in the bowels of the earth at one of the country's many, often little-known, coal mines.

These factors however do not fully account for the relative lack of labour mobility within the salt industry, or indeed why salters do not appear to have attempted to leave it altogether. It is conceivable that such was the effectiveness of the gaze of the grieve's 'eagle eye', allied to the physical isolation of the works, that escape was difficult[177] and saltworkers were instead forced to meekly accept their fate. This view however does not rest easily with the evidence. Where possible, saltworks were sited near to the access points for their markets, near to harbours and in towns, and were frequently focal points for their surrounding communities.[178] It was practically impossible for a salt-grieve, with a variety of managerial functions to fulfil, to keep a close watch on the activities of salters working in enclosed panhouses. What of the salters themselves? How did they feel about and perceive their circumstances? Whilst these questions probably defy all but speculative responses, there are grounds for arguing that in spite of a hostile and potentially degrading legal framework and the demands of a physically-taxing job, the salters in Fife and Tayside maintained a considerable degree of independence and self-esteem, at least within their working environment. As in the case of coal mining, conditions were not sufficiently poor to discourage people from entering the trade voluntarily, as in the case of the aforementioned Walter Smith at Dysart in 1752, or James Frazer and William Gunn, both 'lately in the Shire of Sutherland', who freely became salters at the same works in 1756.[179]

Throughout the period salters managed to sustain a working pattern which allowed time for both leisure and pleasure. Yule and New Year breaks, for example, had long been notable features of the Scottish coal mining and salt making communities, often frowned upon by their employers. The evidence from the eighteenth century presents an inconsistent picture, with production being little affected at some works.

An active desire to continue the tradition clearly continued at Methil, however. In February 1777 George Stirling, a master salter there, was given additional coals 'in consideration ... for his going to work with his pan the 13th Jany last in opposition to the threatenings of his neighbours', whilst another salter's servant, Christopher Jamieson, who worked with him 'when none of the rest would do it' was given an extra shilling.[180] Fast days too were observed and, as late as June 1817, David Greenhill and Alexander Adamson, at Dysart, each lost a day's output for this reason. Pans were also stopped when their operatives went off on personal or family business. In July 1725 John Foord at Bogie ceased work for a day because of a 'death in his famillie', as did a salter at Dysart, in September 1817, on the occasion of his mother's death. As the output from these absences was rarely made up, the decision to forego earnings points to a strong desire on the part of the salters to fulfil what they considered to be their individual and communal responsibilities.

Some elements of paternalism appear to have bonded the saltmaster-salter relationship, although these are always difficult to distinguish from sheer self-interest. Apparently spontaneous and generous actions at times when salters were sick, involuntarily idle or had performed exceptionally well, maintained loyalty. The best might have left, or more likely, become dissatisfied and recalcitrant, if there was too much interference in working patterns and practices which the sub-contracting system had, in part, allowed the salters to establish. Relations between masters and men however were neither wholly one-sided nor free from conflict.

Although the evidence is limited, there is more than a suggestion that during the seventeenth century, and sometimes beyond, the salters were able to exercise some collective influence over the terms of their employment. It is not without significance that the 2nd Earl of Wemyss noted in 1654 that he had agreed terms with 'my whole Salters', with whom he frequently negotiated as a body. For Wemyss this was a difficult period, as clauses in the agreement with his salters were added, strengthened or removed as he sought, over several years, to establish a set of rules which would produce the returns which he wanted and at the same time be acceptable to the salters. He had clearly stated obligations, such as the payment of 'a firlot of meal if they [the pans] stand still in storms or otherwise when they are not cleaning', which were to be taken seriously - as can be seen in the clause which insisted that the salters were not to 'set' (or stop) their work 'till first they show me or my grieves that I have failed in these conditions to them'.[181] Although there is little evidence of formal organisation amongst the salters, it seems quite possible that it was their potential for disruptive action of some sort which evinced the concern amongst the saltmasters to pay them for their salt regularly, either weekly or fortnightly, in cash. Cash payments or quantities of ale were often disbursed when the salters were called upon to carry out additional tasks, although the difference in practice between saltworks has to be stressed. Customary obligations were recognised by many masters at least until the early eighteenth century; for instance at Dysart, in October 1713, six pints of ale were given to the salters when a new set of 'salt firlots', the containers used to measure the quantities of salt they were to be paid for, was being made and measured to the satisfaction of all concerned.[182]

There are signs, however, that as the eighteenth century progressed labour relations within the industry became increasingly bitter. Changing economic conditions forced the saltmasters to throw off the velvet gloves with which they had formerly handled their salters, who in turn conducted an ill-organised struggle to maintain what they may have judged to be long-established rights. Perhaps the best example concerns the question of the salters' 'overplus'. During the half century or so of the industry's rapid expansion from the 1570s, it had become common practice for the salters to retain and sell their 'overplus', that is the salt which they managed to make over and above the quantities they had contracted to sell to their masters for an agreed supply of coal. Generally, the only stipulation was that sales of 'overplus' salt should not take place until after the full amount of contracted salt had been delivered to the girnels. Thereafter the salters could dispose of it as they wished, but at most works at least some of it was purchased by their masters who then sold it themselves. The salter thus had opportunities for private gain, while the fixed quantity he was bound to make beforehand guaranteed that the saltmaster's requirements were met too.

The liberality of this system whereby, according to Thomas Tucker, the masters were 'noe wayes able to arrive ... at what was truely made' by their salters,[183] was subjected to increasing scrutiny as, in the latter decades of the seventeenth century, the saltmasters turned their attentions towards the home market which their 'workemen's salt' had partly supplied. Another concern, which they shared with other contemporaries of their station, was that their employees' earnings should not rise above a certain level[184] and thereby remove the stimulus to hard work at the pan. Serious attention began to be paid to the quantities of salt which a pan could make and how much it was reasonable for a salter to have as an 'overplus'. In March 1727, for instance, when John

Foord and his assistant were able to make 6½ bolls over and above the agreed 'Draught' (the quantity of salt to be delivered for a given amount of coal) of 43½ bolls, and sell the 'overplus' for 19s.6d. (Scots), this was considered to be a 'pretty good fee for ten days service'. Ten bolls, however, was 'far too much' for a week's wages.[185] Three years earlier, at the same works, a salter, who was almost certainly from another work, was paid £6.18s. (Scots) for 'tryall of pan coals ... for 3 bolls 3 firlots of Salt more than Taught', an experiment whose success would undoubtedly have led to attempts being made to raise the 'Draught', with a consequent reduction of the 'overplus'.

An additional restriction upon the salters had been imposed in 1713, when excise duties on Scottish salt, reluctantly agreed to by the Scots at the time of the Union, began to be collected. All salt, no matter how small the quantity, had to be accounted for to the salt officers who were thereafter stationed at each of the country's saltworks. A period of rampant illicit dealing in salt ensued, with salters at every works for which records have survived being fined or imprisoned for attempting to sell salt upon which duty had not been paid. While the total amounts involved may not have been large,[186] for the individual salter the sale of a few bushels of salt - at between 9s. and 12s. (Scots) each in the early eighteenth century - was an important supplement to his income. For salters who had failed to fulfil their 'Draught' and who owed their masters what was termed 'hind' salt, illicit sales were likely to have provided their sole source of cash. The authorities, however, were determined to close the loophole which they perceived to be the major cause of the problem. A clause was inserted into an Act of Parliament, whose primary concern was to alter the duties payable on herring, which declared that from June 1735 it would be illegal for saltmasters to 'pay the Wages of any Servants employed in them [salt works] ... in Salt'. Transgressors could be fined £20.[187]

This was a serious matter for the salters, especially as it coincided with a rapidly growing interest amongst Fife's saltmasters, notably those in the Kirkcaldy district, in salt sales within Scotland, as the last exporting boom began to peter out. The state and the salt proprietors were now united in the cause of removing all opportunities for salters to sell salt privately from the pans. The illicit sale of what was, after 1735, stolen salt continued however, and salters devised ingenious means of deceiving the salt officers. It was alleged that they removed salt from the pan at the 'second water', that is when the pan was being refilled a second time, long before a 'full' pan of salt had been made, but when a perfectly saleable commodity had been produced. The officers' attentions were evaded by the salters keeping the pan 'in a strong boiling state ... and the houses full of steam', even though the evaporation of the water was virtually complete. To guard against this eventuality, officers required to know when fresh water was to be added, and to attend the pan as necessary - which involved unpopular night work. A 'material part of the duty' of a salt officer, the Commissioners of Customs reminded their employees, was to be watchful at all times, 'particularly in the Night time at uncertain hours'.[188] This was not least because salt grieves themselves (perhaps with the connivance of their masters) were not averse, on occasion, to evading the duties and selling salt from the girnels improperly, usually without it having been weighed by the officers. Such may have been the intention of James Young, Sir John Anstruther's agent at St Monance, who was discovered in the 'Granary' (girnel) at 3 o'clock on an August morning in 1776, in spite of the Salt Office's ruling that salt could only be taken out between sunrise and sunset.[189] The salters however did not restrict themselves to a single method of pilfering salt. After having been drawn from the pans, quantities were sometimes set aside 'in a corner of the boiling house, on the pretext that it was foul' and later sold. Small quantities too were removed from the 'drabs' or baskets in which it was taken from the pans to the girnel.

It is not easy to account for the longevity of this activity, which continued through the rest of the period. It can of course be dismissed as a manifestation of the wayward habits of a societal sub-group with strong criminal proclivities. Even in the middle of the seventeenth century, when salters had the right to sell their own 'overplus', they enhanced their incomes by stealing salt. The saltmasters, Tucker wrote, found it difficult to 'keepe theyr ... salt from being embezzled ... through the vileness and unworthinesse of theyre owne workemen'.[190] Such a view would have found much favour with the historian of St Monance who had frowned upon the local salters' boast that the ability to pilfer salt was an 'indispensible qualification' of their trade.[191] Nineteenth and twentieth century attitudes to property however cannot realistically be applied to the centuries which immediately preceded them. In addition, while most of the country's salt officers and watchmen were judged by their superiors to be fairly reliable,[192] there were enough who were prepared to turn a blind eye to the improper habits of the salters, in whose communities they lived as a minority, for theft to remain a socially acceptable occupation. Occasionally the capacity of an officer to carry out his duties was clearly suspect, as in the case of James Petrie, a salt watchman at Usan from 1795, who was 'lame of one leg, which is shrunk ... and thereby rendered unfit for going on board of, or coming ashore from, Vessels', and was considered incapable of earning a livelihood for himself and family by ordinary labour. Petrie however was judged by his superiors to be a capable watchman, although rather oddly, given the nature of his job, he was unable to write. It is also worth noting that there was an unconfirmed suspicion of frauds being committed with the connivance of the officers at Usan in 1796. The officers there were housed in a building which was attached to the home of the saltwork manager.

Even so, it is hard to ignore the possibility that the salters' actions were at least partly influenced by unspoken assumptions about traditional practices and customary rights - similar to the colliers' belief in their right to a free quantity of coal.[193] An example is to be seen in the case of 'Sunday Salt', about whose existence the authorities were apparently unaware until the early nineteenth century. This was a high-quality salt, consisting of large grains which were produced by the slow evaporation of a quantity of water, which the salters left in the pans on Saturday night so that on their return to work on Monday morning they were not required, as otherwise would have been the case, to warm the cold pans slowly and perhaps add 12 hours to their production time. As far as can be ascertained this salt was considered to be their own. It is tempting to believe that this custom had originated in the 1640s, when one consequence of the 'rule of the pious' had been the salters' enforced observation of the Sabbath. As was seen in Chapter 2 there had been considerable reluctance on the part of both the saltmasters and the salters to stop their pans. While Kirk pressure eventually forced the Earl of Wemyss and his near neighbour, Lord St Clair, to order that the pans 'stand' on Sundays from six or seven in the morning, depending on the time of year, 'till 10 at nyt',[194] the salters were still able to enter the pans on the Sabbath outwith the stated hours. Furthermore, they were encouraged to break with the spirit, if not the letter, of both the law and the Kirk's directive, by staying away from the works but 'still keeping fire under the pan'.[195] An allowance of coal for this purpose would have been required and, presumably, provided. Later, in 1660, Wemyss urged that the salters should allow the pans to go cold 'before any person go to Church', but by then the habit of keeping them warm may have become established. By countenancing Sunday work in the 1640s, when Kirkcaldy presbytery had considered 'the multiplied relapses of salters in Sabbath breaking and how frequentlie they ... mocked God with simulating publik repentence', [198] the saltmasters can only have lessened their salters' respect for discipline and authority. Whatever the provenance of the practice of making 'Sunday Salt', salters ceased to benefit from it from 1808, after salt officers became aware of its existence and that it was being carried off 'late in the Sunday evening, or before they [the salters] begin to charge the pan on the Monday morning'.[197]

While salt stealing was generally conducted on an individual basis, it can perhaps be legitimately interpreted as a manifestation of a degree of collective consciousness amongst the salters in that many of them were engaged in this activity, which was clearly against the interests and wishes of both their employers and the state. Thus, as has been noted, five salters from Wemyss saltworks were simultaneously incarcerated in Kirkcaldy jail for selling salt illegally in March 1753. In what was clearly an attempt to break down the salters' solidarity in crime, James Todd, the salt agent at Methil, paid George Stirling an additional coal bonus in 1772, 'as a Mark of his Honble Masters regard for his honesty'.[198] Throughout the following twelve months, Stirling continued to ignore whatever pressures his fellow salters applied and early in 1773 was rewarded with an extra load of coal 'for his punctuality and Honesty ... while the rest did Smuggle'. The other salters were warned that 'they need not Expect or demand any thing whatever in name of winter allowance' until they swore that none of the salt made in their pans would be 'Abstracted, Smuggled or Run'.

Unfortunately, the evidence concerning the theft of salt and its illicit sale in Fife becomes indistinguishable from that of the rest of Scotland from the later eighteenth century. No detailed material about this or any other aspect of the detailed operation of Tayside's saltworks has yet become available. Clandestine activity carried on; indeed there are strong indications that it rose to a new height at the very end of the eighteenth and early years of the nineteenth centuries, as a result of wartime-induced price rises and steep increases in the salt duties which made it a very lucrative activity.[199] It was during this period that the salters at St Monance had been seen 'bursting with much self-complacency that they could hoodwink the guager, and smuggle in his very presence, however vigilant in the discharge of his duty'.[200] Opportunities for 'private sale under a modified duty', however, vanished in 1823 when the salt duties were repealed. Additional cuts in the salters' incomes were enforced by the salt price reductions which lower-priced English imports necessitated. Many faced unemployment as works closed, their skills redundant after the industry finally lost its long-held privileged position within Scottish society.

CHAPTER 4

Conclusion

Between 1570 and 1850 the salt industry in Fife rose from small beginnings and ultimately entered on a course of rapid and unstoppable decline. Tayside's entry to the industry was sudden and short-lived. Lack of quantifiable data, especially for the seventeenth century, makes the task of tracing the industry's progress a hazardous one, but the indications are that in its early stages growth was overwhelmingly dependent upon exports. From the middle of the seventeenth century Fife's saltmasters increasingly turned their attentions to the home market. A second surge of export-based activity occurred in the latter years of the century, and lasted into the eighteenth. After the collapse of the industry's final exporting boom of the 1720s and 1730s, the Scottish market was all that remained for the Fifers, and any growth was dependent upon the rate of population increase. There is no evidence that per capita salt consumption rose. However, to have effected a transfer from the overseas to the home market with only a slight fall in total output between the 1710s and 1790s was a considerable achievement, especially as Fife's saltmasters had formerly relied so heavily upon exports. Nonetheless the region's share of Scottish salt sales did decline, from 52 per cent between 1716-19, to 42 per cent by 1795-98. If Tayside's sales are included the latter figure increases to 46 per cent. The construction of saltworks in Tayside and elsewhere on the north east coast meant that the salt masters in Fife benefited less from the Scottish salt industry's final prosperous phase, from around 1806 until 1823, than did those on the south bank of the Forth or in Ayrshire. Until 1823 though the industry's decline was relative rather than absolute. Salt manufacturing was increasingly overshadowed by the new and rising industries which formed the backbone of Scotland's industrial revolution.

Marketing difficulties there may have been, but the saltmasters did not simply lie back and accept defeat. From the 1660s they combined with most of the rest of Scotland's salt proprietors to demand and achieve protection for their industry from the Scottish Parliament and Privy Council, of which several of their number were prominent members. Their privileges, including rights to restrict the freedom of movement of their workers, were maintained through the Union of 1707, and indeed they retained their commercial monopoly until 1823. Fife's saltmasters were often at the head of eighteenth century campaigns to preserve the status quo whenever suggestions arose that English rock salt should be freely allowed into Scotland. By the late eighteenth century however there was less need for protection. Most of the Forth-side collieries, many of which had formerly been dependent for their survival on sales of small coal to the saltpans, had found alternative and more lucrative outlets, to limeworks, ironworks and the domestic consumer. Salt works had by this time become appendages of collieries. Their continued protection from English competition was simply an anachronism. Economically and politically the tide had turned and was flowing strongly against the Scots saltmasters; the state could no longer justify the existence of a monopoly which denied to the Scottish domestic consumer or industrial user, mainly in the vital chemical sector, a cheaper and superior variety of salt.

Technologically the industry provides a prime example of uneven development, although further investigation will be necessary if the reasons for the variety of practices are to be fully understood. Nothing is currently known about the sizes and types of pan used in Tayside's salt works, nor from where working practices were copied or learned. It has however been clearly demonstrated that salt making was not a job which could be easily or successfully carried on by the novice.

It is to be hoped that this study has partly rescued the salters from the recesses of Scottish history. It is true that the colliers and salters were often treated identically by the law. Further, commonly sharing the same masters it comes as no surprise to find that salters were often subject to the similar restrictions as the colliers. Yet they were a distinct group, not only occupationally but also in that the master salters appear to have achieved a higher status than the colliers, at least within the environment of the colliery-saltwork. In part this may reflect their skill and the individual responsibility inherent in the saltmaking process. Salt sales too played a more important part in the profitability of a colliery-saltwork undertaking in the sixteenth and seventeenth centuries, and beyond, than is sometimes assumed. As a commodity salt was then held in far higher esteem than it was later on. However, generally less buoyant conditions from the middle of the seventeenth century forced the saltmasters to impose a stricter regime than had been the case formerly. As the industry's importance was reduced in the eighteenth century, both as an income earner for the coastal landed estate and nationally, so the salter's status was further eroded. Questions remain, not only about the critical area of salters' earnings, but also about when and where the contractual arrangements, working practices and customs had originated. Certainly the salters appear to have been engaged in a long drawn out struggle to maintain the standards and traditions, which had become established by the early seventeenth century, in the less favourable market conditions and increasingly profit-conscious eighteenth century. More work could usefully be done on the inter-relationships between the collier and salter communities - was there, for example, much inter-marriage between the two groups?

This study then has a number of limitations. Undoubtedly readers will find other flaws. Neverthless it may serve to inspire further research, either at the local level within Fife and Tayside, or, perhaps of greater immediate importance, on the major saltworking areas on the south side of the River Forth, at Cockenzie, Prestonpans or Bo'ness. As has been suggested herein, there are reasons to suppose that the family working unit may have been more important there than in Fife. Similarly, salt voucher material makes it clear that there were many more proprietors, well below the rank of landowner, at Cockenzie and Prestonpans than elsewhere. Indeed it is possible that these were often proprietor-salters, which must rule out the applicability of the concept of the salter-serf in these cases. At the other extreme it requires to be explained why Fife's salters were accommodated in housing which was physically separate from their place of work while their counterparts in Ayrshire lived in an upper floor of the panhouses.[201] Much work remains to be done on the Scottish salt industry; the appearance of this essay may make the task a little more manageable.

NOTES

Abbreviations:

N.L.S. National Library of Scotland
N.S.A. *New Statistical Account*
O.S.A. *Old Statistical Account*
S.R.O. Scottish Record Office

Introduction

1 *Scottish Monthly Magazine,* July 1836, 129.

2 J. Butt, *The Industrial Archaeology of Scotland* (Newton Abbot, 1967), 132; M. Plant, *The Domestic Life of Scotland in the Eighteenth Century* (Edinburgh, 1952), ix, 97-8.

3 Sir G. Mackenzie of Rosehaugh, *Memoirs of the Affairs of Scotland* (Edinburgh, 1821), 171.

4 Quoted in S.G.E. Lythe, *The Economy of Scotland in its European Setting, 1550-1625* (Edinburgh and London, 1960), 51.

5 Sir A. Mitchell (ed.), *MacFarlane's Geographical Collections* (Edinburgh, 1906), 1, 308.

6 Wemyss Castle MSS, Wemyss Castle, Fife, Oncost Book, Dysart Saltworks, 3 May 1817-7 February 1818.

7 T. Tucker, 'Report upon the Settlement of the revenues in Scotland AD 1656', *Miscellany of the Scottish Burgh Records Society* (Edinburgh, 1881), 20.

8 D. Loch, *Essays on the Trade, Commerce, Manufactures and Fisheries of Scotland* (1778) II, 47.

9 S.R.O. Coal Board Records, CB 27/2, Vol 2, R. Bald's Report on the Colliery and Saltworks of Dysart, 13 November 1817.

10 B. Lenman, *From Esk to Tweed* (Glasgow and London, 1975), 32.

11 Derived from calculations based on S.R.O. E 536/1-84, Salt Charge Vouchers, 1713-1798.

12 C.A. Whatley, 'Scottish salt making in the 18th century: a regional survey', *Scottish Industrial History,* 5.2 (1982), 5, 16.

13 J. Roger, *General View of the Agriculture of the County of Angus or Forfar* (Edinburgh, 1794), 16.

14 India Office Records H/730 (Home Misc. Vol. 730), John Leighton to David Scott, 14 June 1796. I am indebted to Mr A. Muir of Dundee College of Technology for this reference.

15 I.A. Adams, 'The Salt Industry of the Forth Basin', *Scottish Geographical Magazine,* 81 (1965), 153-162.

Making Salt: Problems and Processes

16 C.A. Whatley, 'An Early Eighteenth Century Scottish Saltwork: Arran, c. 1710-1735', *Industrial Archaeology Review,* IV, 1982, 95-6.

17 S.R.O. Customs and Excise Records, CE 53/4/4 Customs and Excise, Montrose, Salt Letter Book, 1794-1803.

18 *Acts of the Parliament of Scotland,* V, 419.

19 N.L.S. MS 3088, Bogie Salt Book, 1720-1732.

20 Wemyss Castle MSS, Oncost Book, Dysart Saltwork, 1817/18.

21 For a recent discussion on this topic see J. Rule, *The Experience of Labour in the Eighteenth Century* (1981), 52-57.

22 Wemyss Castle MSS, Diary of David, 2nd Earl of Wemyss, Agreement of Whitsunday 1657. This is a typescript transcript of the original.

23 S.R.O. Rosslyn MSS, GD 164/5/37, Saltwork Accounts, 1752.

24 N.L.S. Bogie Salt Book.

25 S.R.O. GD 164/49/398, 'Account of Smith Wadges for the pan beeting, 1715'.

26 S.R.O. Henderson of Fordel MSS, GD 172/442/9, Account sent by Sir Robert Henderson of Fordel to Hugh Grandiston, 1 April 1755.

27 Adams, 'Salt Industry', 159-60.

28 Tucker, 'Report', 5.

29 Wemyss Castle MSS, Diary.

30 N.L.S. Dundas of Dundas MSS, Adv. MS 80.1.3, Archibald Robertson to George Dundas, 20 January 1720.

31 I.A. Bowman, 'Culross Colliery: a Sixteenth Century Mine', *Industrial Archaeology*, 7 (1970), 368; A. Cochrane, *Description of the Estate and Abbey of Culross* (Edinburgh, 1793), 10.

32 S.R.O. GD 164/5/37 'Acct. Genl Sinclair To Alex^{r} Mawer for the Saltpans after the Great Storme 1752'.

33 N.L.S. Bogie Salt Book.

34 Tucker, 'Report', 5.

35 Wemyss Castle MSS, Dysart Oncost Book, 1817/18.

36 Ibid.

37 S.R.O. GD 164/49/398, 'Acct of Salt Sold and the Money payed to David St Clair Since July 25 1713'.

38 Wemyss Castle MSS, 'Extracts from Wemyss Castle Muniments', Vol. III, Discharge, 'Doctor Hay for Several Wemyss Salters Fines', 1753.

39 Saltworks were 'variously constructed' in England too. See W. Brownrigg, *The Art of Making Common Salt* (1748), 50.

40 'Journal of Henry Kalmeter's Travels in Scotland, 1719-20' (ed. T.C. Smout) in R.H. Campbell (ed.) *Scottish Industrial History: A Miscellany* (Edinburgh, 1978), 41.

41 O.S.A., (Edinburgh, 1978 ed.), 745.

42 S.R.O. Dundonald MSS, GD 233/183/Folder 15, Processes and Papers concerning the sale of Culross, 'Report by Messrs Henderson, Grieve etc., as to the Forest of Culross ... 16 October 1795'.

43 N.S.A., Vol. IX, 187.

44 S.R.O. GD 164/60/504, 'An Account of the Quantities of Salt Made... at the several Salt works upon the Firth of Forth', 1808-1817.

45 N.L.S. Adv. MS 80.1.3. A. Robertson to G. Dundas, 20 January 1727.

46 'Journal of Henry Kalmeter', 44.

47 C.A. Whatley, 'The Finest Place for a Lasting Colliery: Coal Mining Enterprise in Ayrshire c. 1600-1840', *Ayrshire Archaeological and Natural History Society*, 14 (1983), 68-9.

48 'Journal of Henry Kalmeter', 44.

49 S.R.O. GD 172/503/8, Note of pans at St Davids, Whitsunday 1788.

50 S.R.O. CB 27/2, R. Bald's Report on the Colliery ... of Dysart, 1817.

51 *Scots Magazine*, (1804), 66, 760.

52 S.R.O. CB 27/1, R. Bald's Report on Preston Island Colliery and Saltworks, 2 January 1817.

53 'Journal of Henry Kalmeter', 44.

54 *Ibid.*

55 S.R.O. GD 172/1107/34, Salt Agent's copy letter book, 15 March 1845.

56 Adams, 'The Salt Industry', 155.

57 Wemyss Castle MSS, Diary.

58 'Journal of Henry Kalmeter', 41.

59 S.R.O. CB 27/1, R. Bald's Report on Preston Island ... 2 January 1817.

60 Brownrigg, *op. cit.*, 53.

61 University of Edinburgh Library, Gen. 874/IU/29-30, Archibald Cochrane to Professor J. Black, 17 February 1786.

62 Customs and Excise Office, Kirkcaldy, Kirkcaldy Letter Book, Collector to the Board, 24 May 1758.

63 *Scots Magazine*, 66, 762.

64 *Ibid.*

65 Wemyss Castle MSS, Ledger of Salt Accounts, Methil and Wemyss, 1771-76, salt agent to William Stewart, 15 October 1774.

66 S.R.O. CE 53/4/4, Customs and Excise, Montrose, Salt Letter Book, Collector to the Board, 20 July 1796.

67 *Register of the Privy Council of Scotland*, First Series, II, 265-6.

68 S.R.O. GD 164/49/398, 'Reasons to be given in against the valuation of coall and salt works being Casuall rents', n.d. (late seventeenth century). Other authorities claim, more realistically, that there may have been 35 pans in the Kincardine area.

69 N.L.S. Pitfirrane MSS, MS 6456, Account Book, Pitfirrane Coal and Salt, 1703-1746.

70 Wemyss Castle MSS, 'Extracts from Wemyss Castle MSS', Vol. II, List of Forth Salt Proprietors, and numbers of their pans, 1 June 1789.

71 N.S.A. *op. cit.*

Output, Sales and Markets

72 I. Guy, 'The Scottish Export Trade, 1460-1599, from the Exchequer Rolls' (unpublished M.Phil. thesis, University of St Andrews, 1982), 141, 151-9.

73 T.C. Smout, *Scottish Trade on the Eve of the Union, 1660-1707* (Edinburgh and London), 229-30.

74 Guy, 'The Scottish Export Trade', 158.

75 S.G.E. Lythe, 'Scottish Trade with the Baltic, 1550-1650', in J.K. Eastham (ed.) *Economic Essays in Commemoration of the Dundee School of Economics* (1955), 76; J.I. Israel, *The Dutch Republic and the Hispanic World, 1606-1661* (Oxford, 1982), 92, 204, 214-5.

76 *Register of the Privy Council of Scotland,* First Series, II, 407.

77 Cochrane, *Description,* 15.

78 W. Stephen, *The Story of Inverkeithing and Rosyth* (Edinburgh, 1938), 91.

79 A.S. Cunningham, *Inverkeithing, North Queensferry, Rosyth and the Naval Base* (Dunfermline, 1903), 207.

80 Wemyss Castle MSS, Diary.

81 Israel, *The Dutch Republic,* 339.

82 Lythe, 'Scottish Trade'; calculated from the table on page 74.

83 A.S. Cunningham, *Romantic Culross, Torryburn, Carnock, Cairneyhill, Saline and Pitfirrane* (Dunfermline, 1902), 35; S.M. Jack, *Trade and Industry in Tudor and Stuart England* (1977), 88-9; J.U. Nef, *The Rise of the British Coal Industry* (1932), I, 178; Smout, *op. cit.*, 230.

84 W. Stevenson (ed.) *The Presbytrie Booke of Kirkcaldie* (Kirkcaldy, 1900), 240.

85 D. Beveridge, *Culross and Tulliallan* (Edinburgh and London, 1885), I, 242.

86 Wemyss Castle MSS, Diary.

87 D. Woodward, 'Anglo-Scottish commercial policy during the 1660s', *Scottish Historical Review,* LVI (1977), 166, 174.

88 Smout, *Scottish Trade,* 230.

89 T.M. Devine and S.G.E. Lythe, 'The economy of Scotland under James VI: a revision article', *Scot. Hist. Rev.*, L (1977), 92; Tucker, 'Report', 4.

90 Sir A. Mitchell (ed.), *MacFarlane's Geographical Collections* (Edinburgh, 1907), II, 408.

91 Cochrane, *Description,* 15.

92 Tucker, 'Report', 5.

93 *Ibid,* 4.

94 *Register of the Privy Council of Scotland,* Third Series, II, 53.

95 L.B. Taylor (ed.), *Aberdeen Council Letters* (Oxford, 1952), IV, 275.

96 N.L.S. Yester MSS, MS 14493 f.84, 'Proposalls concerning Salt', n.d. (c. 1670).

97 This will be fully discussed in C.A. Whatley, 'Sales of Scottish Marine Salt, c. 1714-1823', *Scottish Economic and Social History* (forthcoming).

98 J. Ellis, 'The Decline and Fall of the Tyneside Salt Industry, 1660-1790: A Re-examination', *Economic History Review,* Second Series, XXXIII (1980), 53.

99 S.R.O. GD 172/118/1, Lease, Burgh of Inverkeithing and David Maither, 1693.

100 Stephen, *Inverkeithing, 92.*

101 Ellis, 'The Decline', 49; W.H. Chaloner, 'Salt in Cheshire, 1600-1870', *Transactions of the Lancashire and Cheshire Antiquarian Society,* 71 (1961), 67.

102 I. Bisset, 'Commercial and Economic Developments in Late 17th Century Dundee' (unpublished M.A. dissertation, Department of Modern History, University of Dundee, 1978), 14.

103 N.L.S. Ch. 613, Account, Alexander Williamson and Sir John Wemyss, 1702

104 Virtually all the figures which follow are derived from S.R.O. E 536/1-84, Salt Charge Vouchers, 1713-1798.

105 Wemyss Castle MSS, 'Extracts', II, Earl of Wemyss to Alexander Christie, 19 January 1744.

106 Whatley, 'Sales'; see also *Scots Magazine,* op. cit., 817. Further information is available in P.L. Payne, 'The Halbeath Colliery and Saltworks, 1785-1791', in A. Slaven and D.H. Aldcroft (eds.), *Business, Banking and Urban History* (Edinburgh, 1982), 17-18.

107 A.R.B. Haldane, '*The Great Fishmonger of the Tay: John Richardson of Perth and Pitfour*', Abertay Historical Society, 21 (Dundee, 1981), 19.

108 N.L.S. Adv. MS 80.1.4 George Beattie to Robert Dundas, 5 February 1740.

109 S.R.O. CE 67/1/36, Salt Letter Books, Alloa, Collector to the Board, 6 May 1796.

110 Wemyss Castle MSS, Ledger, 1771-6, salt agent to Alexander McKenzie, 27 May 1771.

111 *Ibid,* salt agent to George Jack, 30 December 1775.

112 S.R.O. GD 172/1107/34, Salt Agent's copy letter book, 1 May 1847.

113 *Scots Magazine, op. cit.,* 761.

114 *Ibid.,* 760.

115 Calculated from sales data in Wemyss and Methil Salt Accounts Ledger, 1771-76.

116 Wemyss Castle MSS, Ledger, salt agent to William Marshall, 15 September 1771.

117 *Scots Magazine, op. cit.,* 761; Kirkcaldy Customs and Excise Office, 43/4/4, 'Letters to The Hon. Commisioners of the Salt Duty and ors ... commencing 22 June 1772', Collector to the Board, 18 May 1775.

118 Wemyss Castle MSS, 'Extracts', II, John Turpie to salt agent, 26 December 1796.

119 Loch, *Essays,* 47.

120 Adams, 'The Salt Industry', 159-60.

121 S.R.O. GD 233/183/Folder 15, 'Report', 16 October 1795.

122 Kirkcaldy Customs and Excise Office, Letter Book, Collector to the Board, 30 January 1758.

123 *Ibid.,* 8 January 1766.

124 Wemyss Castle MSS, Ledger, salt agent to James Ramsay, 1 March 1771.

125 Kirkcaldy Customs and Excise Office, Letter Book, Collector to the Board, 27 March 1758, 7 July 1772.

126 Whatley, 'Sales'.

127 S.R.O. GD 172/503/2, James Pinkerton to Sir John Henderson, Bart., 28 March 1788; see too R. Douglas, 'Coal-mining in Fife in the second half of the eighteenth century', in G.W.S. Barrow (ed.), *The Scottish Tradition* (Edinburgh, 1974), 216.

128 N.L.S. Grange MSS, MS 5381 15/2, John Grieve to William Cadell, 26 December 1775.

129 B.F. Duckham, *A History of the Scottish Coal Industry, Vol. I 1700-1815* (Newton Abbot, 1970), 233.

130 J. Girvin, *The Impolicy of Prohibiting the Exportation of Rock Salt From England to Scotland* (London, 1799), 32-3.

131 Montrose Burgh Records, Minute Book of Montrose Town Council, 14 May 1794.

132 S.R.O. CE 53/4/4, Montrose, Salt Letter Book, 20 June 1794.

133 Roger, *General View*, 16.

134 S.R.O. E 536/81, Salt Charge Vouchers, April 1794-April 1795.

135 S.R.O. CE 53/4/4, Montrose, Salt Letter Book, 1 February 1796.

136 *Ibid.*, 20 July 1796.

137 Whatley, 'Scottish Salt Making', 16.

138 S.R.O. GD 164/Box 76/802/1, Messrs Fleming, Rutherford and ors. to Sir James St Clair Erskine of Sinclair, 12 December 1795.

139 Whatley, 'Sales'.

140 Only passing reference to this works can be found; see, for example, S.G.E. Lythe and J. Butt, *An Economic History of Scotland 1100-1939* (Glasgow and London, 1975), 41.

141 S.R.O. CB 27/1, R. Bald's Report, 2 January 1813.

142 G. Hay, *History of Arbroath* (Arbroath, 1899), 375; J.M. McBain, *Arbroath: Past and Present* (Arbroath, 1887), 139.

143 S.B. Calder, 'The Industrial Archaeology of Sutherland' (unpublished M. Litt. thesis University of Strathclyde, 1974), 2 Vols., I, 123.

144 R.J. Adam (ed.), *Papers on Sutherland Estate Management, 1802-1816* (1972), II, Marchioness of Stafford to Marquis of Stafford, 11 July 1814.

145 Calder, 'The Industrial Archaeology of Sutherland', I, 137.

146 For a useful description of the qualities of the main varieties of salt available in the early nineteenth century see 'Salt and the Salt Trade' in the *Scottish Monthly Magazine, op. cit.*, 129-139.

147 S.R.O. GD 164/45/348, Account of Receipts and Expenditure, Dysart Colliery and Saltworks, 19 May 1820-12 May 1826.

148 *N.S.A.* IX, 397.

149 S.R.O. GD 172/862, Fordel Colliery (and Saltwork) Returns, 1823-1843.

150 *N.S.A.*, IX, 187; E. Simpson, *Dalgety - the story of a parish* (Dalgety, 1980), 29.

151 S.R.O. GD 172/1107/34, Salt Agent's copy letter Book, 14 November 1846.

152 *Ibid.*, 21 March 1844, 12, 19, 26 October 1846.

153 Simpson, *Dalgety*, 29.

154 *The Fifeshire Advertiser*, 1 January 1853.

155 S. Chester, L. Foster, L. Hughes, *A Short History of the Villages: Charlestown, Limekilns and Pattiesmuir* (Charlestown, 1979), 20.

Salter Serfs

156 T.C. Smout, *A History of the Scottish People, 1560-1830* (1981 ed.), 168-170, 403-7; W. Ferguson, *Scotland: 1689 to the Present* (Edinburgh, 1978 ed.), 93.

157 Smout, *A History,* 169-170.
158 Beveridge, *Culross,* II, 245-6.
159 A. and N.L. Clow, *The Chemical Revolution* (1952), 55.
160 *Register of the Privy Council of Scotland,* First Series, VII, 434; *Acts of the Parliament of Scotland,* V, 419, 'Acte anente Coallheuares'; the best modern account of the legal aspects of serfdom is B.F. Duckham, 'Serfdom in Eighteenth Century Scotland', *History,* LIV (1969), 178-197, although it deals solely with the experience of the colliers.
161 N.L.S. MS 3088, Bogie Salt Book, 1720-1732.
162 Wemyss Castle MSS, Diary.
163 S.R.O. GD 164/16/124/1, Agreement between Walter Smith and David Beatson, 24 February 1752.
164 N.L.S. MS 3720, Journal of the Management of the Coall and Salt Works of Prestongrange, from 2 April 1748.
165 *Ibid.,* for example.
166 Wemyss Castle MSS, Ledger of Salt Accounts, 1771-76.
167 *N.L.S.* Ch. 651, Petition of William Foord, 11 October 1732.
168 B. Didsbury, 'Cheshire Saltworkers', in R. Samuel (ed.) *Miners, Quarrymen and Saltworkers* (1977), 147.
169 Sir W. Fraser (ed.) *Memorials of the Family of Wemyss of Wemyss* (1888), III, 203.
170 W.R. Scott (ed.) *The Records of a Scottish Cloth Manufactory at New Mills, Haddingtonshire, 1681-1703* (Edinburgh, 1905), xiv, 37; Smout, *A History,* 169; G. Donaldson, *Scotland: The Shaping of a Nation* (Newton Abbot, 1980), 232.
171 Didsbury, 'Cheshire Saltworkers', 142-3.
172 Duckham, 'Serfdom', 181.
173 *Register of the Privy Council of Scotland,* Second Series, VI, 329.
174 N.L.S. Ch. 654, Obligation, Peter Pugston, Salter to Sir James Wemyss of Bogie, 1757.
175 S.R.O. GD 172/442/5, Tack, Sir Robert Henderson and John Henderson, merchant in Inverkeithing, 1754.
176 N.L.S. Ch. 649, 'Instrument of the Laird of Bogie Agt. John Finlay and others, 1705'.
177 Smout, *A History,* 169.
178 C.A. Whatley, 'An Offshoot of the Coal Mining Industry? Scotland's Salt Industry, c. 1660-1823', unpublished paper read to a meeting of The Economic and Social History Society of Scotland held in the University of Dundee, March 1983.
179 S.R.O. GD 164/16/124, Obligement, James Frazer and William Gunn to General St Clair, 1756.
180 Wemyss Castle MSS, Ledger of Salt Accounts, 7 February 1777.
181 Wemyss Castle MSS, Diary.
182 S.R.O. GD 164/49/398/10, Accounts, Dysart Salt Works, 1713-1715.

183 Tucker, 'Report', 7.

184 R.W. Malcolmson, *Life and Labour in England, 1700-1780* (1981), 1-17; Rule, *The Experience of Labour,* 53-4.

185 N.L.S. Adv. MS 80.1.3, Pat. Maxton to George Dundas, 20 and 30 March 1727.

186 Whatley, 'Sales'.

187 8 Geo. II c. 12 1735, 'An Act for Granting and Continuing the Duties upon Salt, and upon Red and White Herrings....'

188 Kirkcaldy Customs and Excise Office, Salt Letter Book, Commissioners to the Anstruther Collection, 25 August 1772.

189 *Ibid.*, Collector to Board, 9 September 1776.

190 Tucker, 'Report', 7.

191 J. Jack, *An Historical Account of St Monance, Fifeshire* (Edinburgh, 1844), 147.

192 This is the impression given in two sources; Signet Library MS 106: 59. An Account of the Officers of the Customs and Salt Duty at the Out Ports in Scotland ... as they stood on the Establishment at the 5th January 1755; all ten officers in the Montrose/Usan collection were considered to be 'capable' in 1798. Their ages ranged from 37 to 58 years, with the mean being 50.5. S.R.O. CE 53/4/4, Montrose, Salt Letter Book, 5 July 1798.

193 Rule, *The Experience of Labour,* 125-134; R. Houston, 'Coal, class and culture: labour relations in a Scottish mining community, 1650-1750', *Social History,* 8 (1983), 8-9.

194 Stevenson, *The Presbytrie Book,* 242.

195 Wemyss Castle MSS, Diary.

196 Stevenson, *The Presbytrie Book,* 304.

197 Kirkcaldy Customs and Excise Office, 63/6/4, Excise General Letter Book, General Letter to all Collectors, 1 September 1808.

198 Wemyss Castle MSS, Ledger of Salt Accounts, 24 February 1772.

199 S.R.O. CE 8/5 Scottish Excise Board, Letter Books, 1798-1800, Board to H.M. Treasury, 9 April 1799; Whatley, 'Sales'

200 Jack, *An Historical Account,* 147.

Conclusion

201 Signet Library, Court of Session Papers, CSP 132; 13, William Orr v. the Earl of Eglinton and ors., 1764, answers for William Orr; S.R.O. RHP 203/1, P. Fleming's Plan of Saltcoats, c. 1798.

APPENDIX 1

Saltwork Data, 1716-1719 (Fife)

Name of Saltwork	Proprietor	Average Annual Sales, 1716-19 (bushels)*	Proportion of Total (%)	Rank
Aberdour	William Wemyss	1,000	0.75	15
Craigflower	Lord Colville	4,300	3.25	11
Dysart	Lord St Clair	23,100	17.39	2
Inverkeithing	Lady Pittencrieff	3,600	2.71	12
Kincardine	Col. John Erskine	8,100	6.10	6
Kirkcaldy	Sir John Wemyss	9,500	7.15	5
Kirkcaldy Links	William Robertson	5,300	3.99	8
Leven	Daniel Peck & Coy.	4,700	3.54	9
Leven Links	Alex. Gibson of Durie	2,400	1.81	14
Limekilns	Sir Peter Halkett	2,600	1.96	13
Methil	Earl of Wemyss	22,900	17.24	3
Newpans	Col. John Erskine	6,100	4.59	7
Torryburn	Col. John Erskine	11,400	8.58	4
Valefield	Brig. Preston	4,500	3.39	10
Wemyss	Earl of Wemyss	23,300	17.55	1
Totals		**132,800**	**100.00**	

Source: derived from S.R.O. E 536/3-5, Salt Charge Vouchers, 1716-19.

APPENDIX 2

Saltwork Data, 1795-98

Name of Saltwork	Proprietor or lessee	(Fife and Tayside) Average Annual Sales, 1795-98 (Bushels)*	Proportion of Total (%)	Rank
Craigflower	Alexander Dickie	7,400	5.48	9
Dundee	Dundee Glasswork Coy.	800	0.62	15
Dysart	John Veitch	10,500	7.77	5
Inverkeithing	John Campbell and Coy.	15,800	11.70	3
Kincardine	Lord Cochrane	300	0.22	16
Kirkcaldy	Nairn, Anderson and Coy.	6,600	4.89	10
Leven	James Christie	9,500	7.03	7
Limekilns	John Burt	1,000	0.74	14
Methil	William Wemyss	22,600	16.73	1
Montrose	James Dickson	8,200	6.07	8
St Davids	Sir John Henderson	15,300	10.36	4
St Mungoes	Lord Cochrane	1,700	1.25	13
St Philips	Newark Coal Coy.	17,200	7.70	6
Torryburn	Messrs Parker & Thomson	2,000	1.48	12
Usan	David Scott	3,200	2.37	11
Wemyss	William Wemyss	21,100	15.62	2
Totals		**143,200**	**100.00**	

Source: derived from S.R.O. E 536/82-84, Salt Charge Vouchers, 1795-98.

* rounded to nearest hundred.